THE
NORTH SEA

GREENPEACE
THE·SEAS·OF·EUROPE

THE
NORTH SEA

Malcolm MacGarvin

C&B

First published in Great Britain in 1990
by Collins & Brown Limited
Mercury House
195 Knightsbridge
London SW7 1RE

Copyright © Collins & Brown 1990
Text copyright © Greenpeace Communications Limited 1990

All rights reserved. No part of this publication may be reproduced, stored in a retrieval system, or transmitted in any form or by any means electronic, mechanical, photocopying, recording or otherwise, without the prior, written permission of the copyright owner.

A CIP catalogue record for this book is available from the British Library

ISBN 1 85585 005 2

GREENPEACE BOOKS
Series Co-ordinator: Kieran Fogarty
Editor: Lesley Riley
Main text: Malcolm MacGarvin
Greenpeace: A Decade of Campaigns by Peter Brookesmith

COLLINS & BROWN
General Editor: Gabrielle Townsend
Art Director: Roger Bristow
Design: Bob Gordon

Filmset by The Setting Studio, Newcastle upon Tyne
Reproduction by Scantrans, Singapore
Printed and bound in Italy by New Interlitho, Milan

This book is printed on a low-chlorine paper, called Silverblade, produced by the Swedish paper manufacturers Silverdalen. This represented the least environmentally damaging option, given our requirements. An increasing number of paper mills are now producing chlorine-free paper, and it is hoped that such paper will become available in a wide range of weights in the future.

CONTENTS

INTRODUCTION	6
THE HIDDEN WEALTH	8
THE HUMAN TIDE	10

PART 1
THE PRESENT AND THE PAST	14
THE TEEMING SEA	28

PART 2
LOST HABITATS	62
FISHING	70
NUTRIENTS	80
POLLUTION	90
GAS AND OIL	104
SHIPPING	112

PART 3
GREENPEACE	120

PART 4
POLITICS	132
INDEX	141
ACKNOWLEDGEMENTS	144

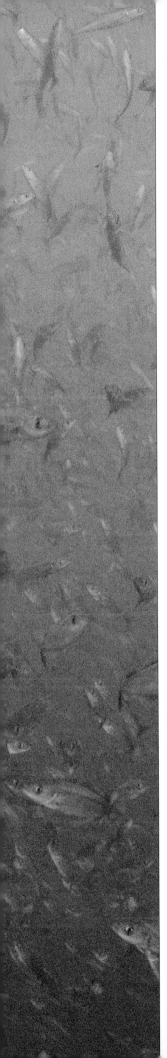

INTRODUCTION

ONE OF THE FIRST REACTIONS of those who come into more than casual contact with the North Sea is astonishment at the profusion of its wildlife. Herring shoals of hundreds of thousands of fish can be commonplace. The animals they feed on are the most abundant creatures in the world, so numerous that if they were humans the entire population of Europe would be crammed into a few tens of cubic metres of the North Sea. Closer to shore, the coastal wetlands rival the tropical rainforests in the sheer weight of living plants and animals that thrive in their sheltered conditions.

This great multitude is intimately associated with the land that surrounds the North Sea. The rivers around its shores naturally supply the coastal waters with large amounts of nutrients, which are the source of all life. Yet this proximity to land has also come close to being the sea's undoing. Because of our changing way of life, the rivers now discharge too many nutrients into the sea; massive amounts of pollutants now enter its waters; and a huge number of fish are pulled out every year, before they have even had the chance to breed. In the most extreme cases, parts of the sea itself have been 'reclaimed'.

All too often when talking about this amazing sea we are forced to use the past tense. Yet the message need not be one of despair. It is true that we are destroying an extraordinary world we scarcely know, almost without thinking. Yet there are many ways to stop this happening. The biggest obstacle now is not a lack of ideas, nor the need for more research, but inertia. This has to be overcome. In all conscience, we cannot put off action until tomorrow. It is our choice: we can continue to cause the life in the North Sea to wither away, or do our best to help bring about its restoration.

Seething profusion
A shoal of predatory saithe, one species among many that can thrive in the nurturing waters of the North Sea. They swim above a plumose anemone, a representative of another abundant group of North Sea wildlife, which acquires its food by filtering organic particles from the water.

THE HIDDEN WEALTH

The impression that many people take away from the North Sea is one of wild empty seascapes of waves and cloud. The sea hides its most remarkable treasure – the wildlife – tantalizingly rolling back the fringe of its mantle twice a day with the tides to give just a hint of the stunning diversity that exists beneath the waves.

The exotic colours and shapes of the life in the North Sea are, quite simply, astonishing. The luxuriant seaweeds, velvet anemones and ethereal jellyfish might more usually be associated with a tropical sea than one set in the heart of industrial Europe – even the worms and slugs have their colourful side. Birds and seals, too, so familiar on land, become different creatures when they enter the water.

Drifting in the currents
This jellyfish (*Rhizostoma pulmo*, above) is one of the largest members of the North Sea plankton, a teeming mass of plants and animals, of astronomical numbers, that drift in the surface waters at the whim of the weather and the currents.

The concealing waves
The tide returns (right) to claim back its own. The plants and animals that live on a rocky coastline are extraordinarily diverse and beautiful. The seaweeds thrive in the bright light of the shallow water and provide a shelter for a community of crabs, shellfish and other creatures.

THE HIDDEN WEALTH

Marine hare
The sea hare (*Aplysia punctata*, left), actually a marine slug, grazes on the lush vegetation that grows in profusion along the rocky coastline. Many small animals, such as sea slugs, worms, shellfish, crabs and starfish, move about the seabed, together making up a key component of the ecosystem.

Masters of air and sea
The gannets (*Sula bassana*, right) have total mastery of the air and water in which they lead their lives. The gannet is just one of the many seabirds for whom the North Sea is an important breeding and feeding ground.

Plush velvet
Beadlet anemones (*Actinia equina*, above) retract their feeding tentacles when the tide is out, but they rapidly extend them again when covered once more by water.

A new generation
A harbour seal (*Phoca vitulina*, left) suckling her young. To many people, seals have become a gauge of the health – and sickness – of the North Sea.

Flying fish
Guillemots (*Uria aalge*, right) and other auks are more agile flying through water than they are in the air.

THE HUMAN TIDE

Over the centuries the fate of many people has been dependent on the North Sea, and this continues to the present day. Once, the main interest was the natural harvest from the sea: fish for the table and seaweed for the crops. From fishing sprang up a seafaring tradition, then a huge shipbuilding industry developed – and yet another group of people became bound up with the sea.

But the tide of human events turned: seafaring and shipbuilding went into decline, the latter reviving briefly during the boom years of oil and gas exploration. And the offshore industry meant that yet another generation found itself dealing with the North Sea. This ebb and flow will continue – but we must learn to use the sea's resources without destroying the sea itself.

The bounty of the sea
Once harvested for use as a natural fertilizer (above), seaweed is now increasingly used in health products, and its gathering may again become a common sight.

Jobs on the doorstep
The homes of shipbuilders (left), dwarfed by the product of their craftsmanship. What was once a thriving industry is at present in a deep recession.

End of the season
A solitary child leaves the promenade at an East Anglian resort (below). The North Sea holiday industry was enormous until the lure of warmer climes and the package holiday intervened. Yet with greater awareness of the North Sea, people may return to its coasts. If and when they do, they will bring both economic benefits to the region and new problems for its environment.

Caught in a trap
A Scottish salmon fisherman attends to his catch (right). All areas of the fishing industry have suffered immense upheavals in the past 30 years as both the fish and the jobs have disappeared.

The oil rush
The discovery of North Sea gas and oil (below) was a godsend for many North Sea states. It created a thriving industry in the middle of the North Sea, and employment in remote areas. Pay can be high but life on an extraction platform is often rough, dirty, cold and dangerous.

The human effect
A seabird lies dead in a pool of oil (above). In our rush to consume the resources of the sea, we have tended to ignore the cost to the natural world. As the pollution of the North Sea has grown, hundreds of thousands of birds have paid for our carelessness with their lives. There is still much we can do to prevent this happening.

PART 1

The North Sea stretches from the rugged coastline of Brittany in the south to the deep Norwegian fjords in the north. Between these shores the coast has many forms, ranging from the wetlands of the shallow Wadden Sea and the huge Rhine delta in its central regions, to the cliffs and rocky shores at its extremities in France, England, Scotland, Denmark and Norway. Away from the coastline, the sea varies in depth, from only a few tens of metres in the southern North Sea, to well over 200 metres in the north.

The North Sea is an ancient sea, with an unbroken presence stretching back 350 million years. During this time it has baked under a tropical sun and been buried deep under ice. It has seen life emerge onto land, mountains raised up and thrown down, and the opening, to the west, of a new ocean – the Atlantic.

The Atlantic now helps to make the North Sea into an extraordinarily diverse ecosystem, by introducing warm water from southern latitudes. This, together with the North Sea's large tides and complex currents, the nutrients that flow naturally into its shallow coastal waters from the surrounding land, and the huge variety of physical habitats, has attracted a stunning profusion of marine life – ranging from fish, birds, whales, dolphins and seals to the many thousands of species of plants and smaller animals.

The spectacular North Sea
The space shuttle Challenger took this photograph on an evening in early November in 1985, from a position north of the Netherlands, looking south west along the length of the continental coastline of the southern North Sea. In the foreground, just beyond the cloud, can be seen the hair's-breadth line of the dam that separates the freshwater IJsselmeer from the Wadden Sea. Further down the coast is the huge Rhine delta marking the Dutch–Belgian border. From there the coast curves to Cap Gris Nez, where just a few kilometres of sea separate the mainland from southeast England. In the distance the Seine catches the sun and the English Channel stretches westwards beyond Cap de la Hague.

THE PRESENT AND THE PAST

THE NORTH SEA is one of the largest coastal seas in the world. An integral part of the European land mass, it is virtually surrounded by land on three sides and, on the fourth, it is blocked by a group of islands, the Orkneys and Shetlands, standing sentinel to the north.

The sea itself consists of a roughly square-sided bay, with the Elbe in the southeast corner, the Wash in the southwest. From the south east, between England and Belgium, a narrow passage called the Southern Bight runs down to a 35-kilometre-wide strait between Dover and Calais; from here the long arm of the English Channel extends westwards to the Atlantic Ocean. On its eastern side, the bay is connected to the Baltic via the Skagerrak, a deep channel between Norway and Denmark, and the shallow Kattegat, which flows between Denmark and Sweden. The North Sea is itself shallow, typically only 25–55 metres deep in the south, falling to 100–200 metres in the north. Finally the sea floor dips into the ocean abyss along a line that runs north east from the Scottish Highlands towards the Norwegian coast and on towards the Arctic Circle.

Tides and currents

The British Isles shield the North Sea from the force of the Atlantic Ocean. But indirectly the Atlantic does have at least one significant effect. The ocean currents bring warm water from the far south and transfer some of their heat to the prevailing westerly winds passing across them. This makes the winter of

Features of the sea
The map shows the North Sea, and the different regions within it, its major rivers and its drainage basin. Water flows into the sea from the north, swirling slowly in an anti-clockwise direction, and from the west, along the English Channel. According to international regulations the North Sea includes the English Channel and extends east as far as the tip of Denmark. This book extends the boundary east, to the southern tip of Sweden.

If the sea ran dry
The water of the North Sea is very turbid, obscuring the seabed from view. However, this radar satellite image (below) penetrates the gloom to show the shape and form of the seabed off the east coast of England. (North is towards the top left of the photograph.) Mud banks and submarine cliffs are clearly visible.

THE PRESENT AND THE PAST

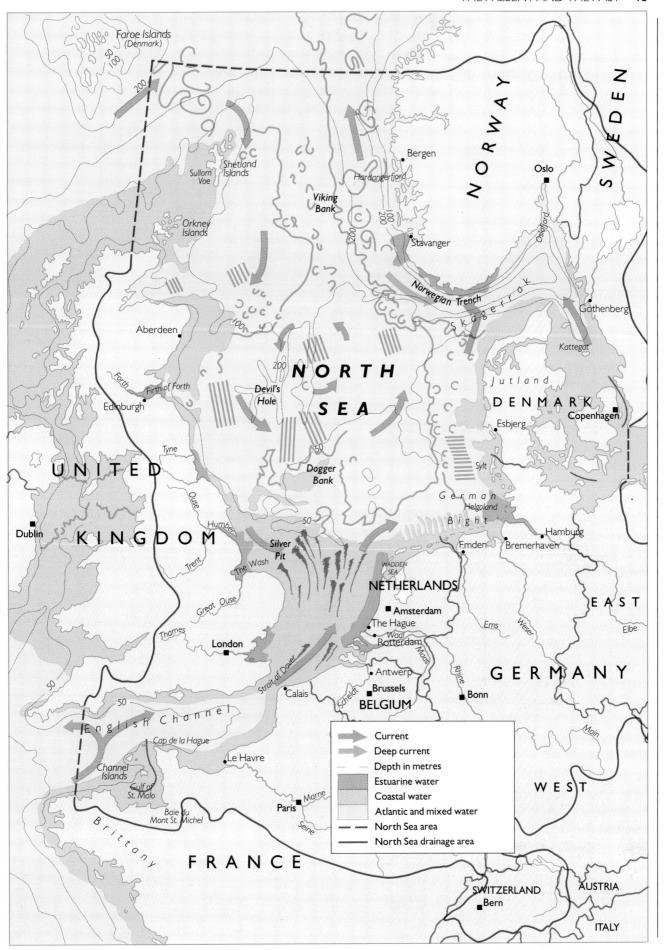

16 THE GREENPEACE BOOK OF THE NORTH SEA

Two seas in one
This satellite photograph of the North Sea (right), taken in early spring 1982, has had the colours enhanced by computer to bring out the physical features of the sea. The very distinct, lighter buff-coloured water along the coasts and in the southern region contains large amounts of sediments. This water, and all that it contains, remains largely isolated from the oceanic water, shown here as dark brown, which enters from the north. Within these two large masses of water many smaller swirls, eddies and boundaries can be seen.

The tidal wave
The map (right) shows spring tides in the North Sea, ranging in height from zero to over 11 metres. Tides may be 30 percent greater than these when they are backed up by a gale (above).

- less than 3m
- 3-6m
- 6-9m
- over 9m

The shifting seabed
The nature of the seabed (left) depends on the direction and speed of the currents, which may heap up sand and gravel, expose surface rock, or shift material along the coast (below).

- mud
- rock
- mud, sand and gravel
- sand

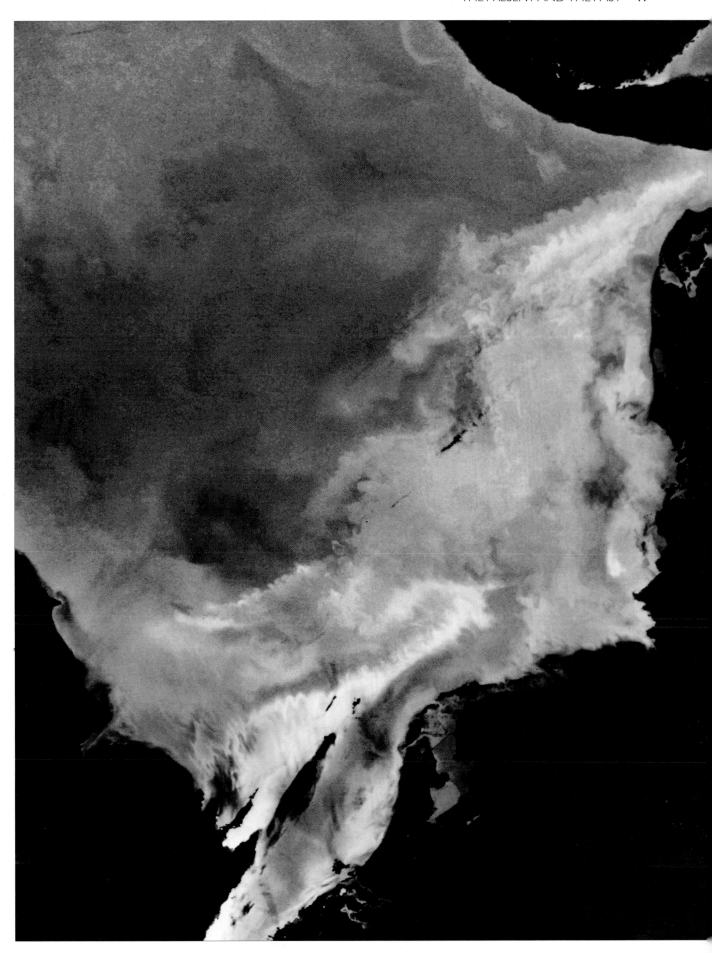

northwestern Europe much warmer. As a result, while other coastal seas of the same latitude, such as Hudson Bay and the Sea of Labrador in the northwest Atlantic, and the Sea of Okhotsk in the Pacific, are partly or completely frozen over in winter, the North Sea typically remains above freezing, at 5°–10°C.

Tides are the most obvious influence on the movement of water in the North Sea. The rising tide from the Atlantic splits into two around the British Isles, entering the North Sea from both the south west and the north. Because it takes different lengths of time for these two tidal waves to reach any one point in the North Sea, they may either work in conjunction, or largely cancel each other out. As a result, the average tidal range – the difference between low and high water – in the North Sea may be less than a metre on much of the Norwegian coast, up to 6 metres on the east coast of Britain, and 11 metres in the Baie du Mont St Michel in northern France.

The tidal motion is dramatic, but the forces that dominate the long-term movement of water in the North Sea are the prevailing currents, which flow in a generally anti-clockwise direction, with quiet patches in the central and northern areas. Regional variations within the North Sea make the pattern of circulation more complicated. There are distinct fronts between different bodies of water, sometimes with interlinking boundaries known as coastal flames, along with huge swirls and other water features stretching over 75 kilometres or more, and visible only from space. The English Channel has its own, often conflicting, set of currents, which vary from season to season. The Skagerrak and Kattegat have a surface current of brackish water going from the Baltic to the North Sea – balanced by a flow of deeper, saltier water from west to east. Near the coasts, water currents are particularly complex. Satellite images have shown that a distinct band of water can remain trapped close to the shore, even when there is a wind blowing from land to sea. Similarly, the flow to and from estuaries is far more involved than was once imagined.

The form of the seabed – its peaks and valleys, ridges and channels – is shaped to some extent by the strength of tides and currents. Where the currents are strong, sediments are moved and rock may be exposed. But for the most part the currents are weaker, and the seabed is covered by sand, mud or gravel. In the southern North Sea and English Channel there are sandbanks and dune fields, with a particularly large gravel bed in mid Channel. In the deeper parts of the northern North Sea such as the Norwegian Trench – a channel running off the coast of Norway – a thick layer of mud covers the seabed.

However, present events only partly account for the structure of the seabed. Most of the sediments in the North Sea were actually deposited in the ice ages; many of the sand banks are relics of earlier patterns of currents; even the rocks themselves are the ghosts of previous incarnations of the North Sea.

INCARNATIONS

Land masses are not fixed to the surface of the globe. Instead, over time, they drift between the equator and the poles, colliding with one another and squeezing seas into oblivion, only to drift apart again millions of years later. Sea levels rise and fall, immersing regions and then draining away again to leave dry land. In this shifting world the North Sea has had a longer existence than most other seas; the North Atlantic is youthful by comparison. The present North Sea is in fact only the latest form of a sea whose history lies entombed in many thousands of metres of rock beneath and around it. Still moving, still changing, its earlier incarnations continue to influence events.

In the Carboniferous period, 350 million years ago, plants and animals were only just beginning to invade the land. At that time, the land that was to become the North Sea was situated about two degrees south of the equator – at about the

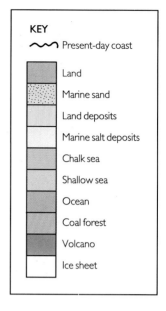

KEY
- Present-day coast
- Land
- Marine sand
- Land deposits
- Marine salt deposits
- Chalk sea
- Shallow sea
- Ocean
- Coal forest
- Volcano
- Ice sheet

Six faces of the sea
The North Sea has changed its shape and size many times in the 350 million years of its existence. From a Carboniferous coal swamp, it became a landlocked sea in the Permian era, and a mass of deltas in the Jurassic. The seas then flooded back to cover most of northwest Europe in a chalk sea during the Cretaceous. By the time of the Eocene period, the North Sea had begun to adopt its present form. This process was interrupted by four ice ages and the sea most recently emerged from the ice 13,000 years ago.

THE PRESENT AND THE PAST

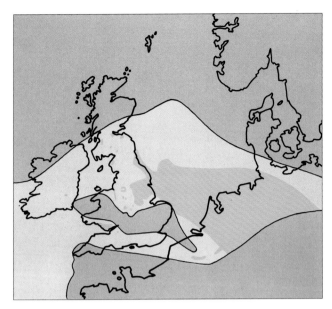
Mid Carboniferous era (300 million years ago)

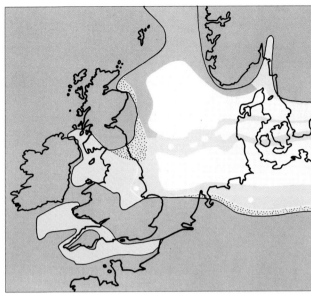
Late Permian era (240 million years ago)

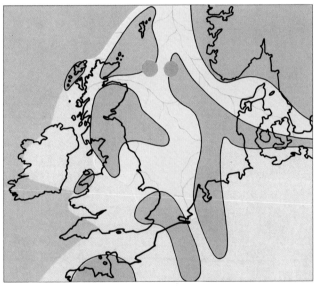
Mid Jurassic era (168 million years ago)

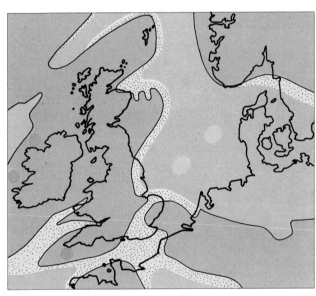
Early Cretaceous era (120 million years ago)

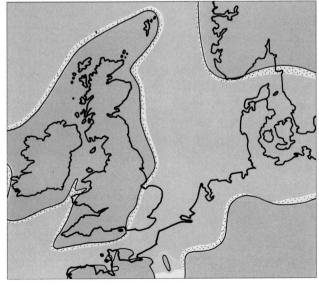
Eocene era (45 million years ago)

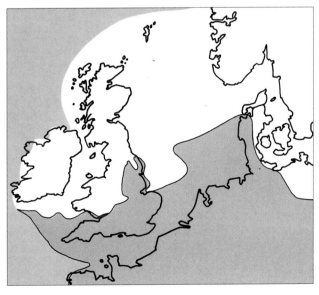
Last ice age (20,000 years ago)

same position as modern Brazilian rainforests. A tropical swamp stretched across the area. Giant ferns and horsetails reared from the shallow water. When they fell, they did not decay, but were gradually compressed by the sedimentary mud that accumulated on top of them, eventually forming enormous coal seams that stretched across the North Sea region.

The coal forests died, and the Carboniferous gave way to the Permian era. By 240 million years ago, the rock under what is now the North Sea had begun to subside – a subsidence that is still in process. What was once a swamp had become a sea, landlocked except for a narrow channel to the north. This was the true geological birth of the North Sea.

By this time, the whole region had shifted to about 20 degrees north of the equator. Its new climate was similar to that of the present-day Persian Gulf. In the intense heat, water was continuously evaporating and being replenished, in the process leaving vast deposits of mineral salts. The richness of the deposits, and the discovery of the uses to which the salts could be put, led to the birth of the chemical industry in the late 19th century. In the mid 20th century, geologists found that the salt deposits held another major resource. Over millions of years, the salts had pushed their way through the layers of rock that had formed above them, creating channels for the escape of gas from the coal seams that lay deeper down. The places where the gas collected, sealed by overlying layers of salt, are today's North Sea gas fields.

As the European land mass continued its slow slide to the north, the seas resurged, to submerge nearly everything except what is now Scandinavia, the Highlands of Scotland, and Brittany. But then, about 168 million years ago, the land again rose slightly, turning the North Sea into an intricate maze of deltas. Many of the marine deposits created during this Jurassic age were rich in organic material, thought to be the source of the oil in the northern North Sea; like gas, the oil was trapped by subsequent impervious sedimentary deposits.

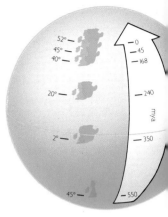

A dance through time
The North Sea started life near the equator. Since then it has made an irregular passage north.

Ancestral remains
The rocks beneath the North Sea are the remnants of its ancestors. This cross section shows the remains of the coal forests, the Permian salt sea, the Jurassic marine limestone and the chalk sea. The rocks in the north have sunk dramatically, the hole being filled by much more recent sediments. Gas from the coal measures (on the extreme left), and oil from the Jurassic age have accumulated beneath impervious rock.

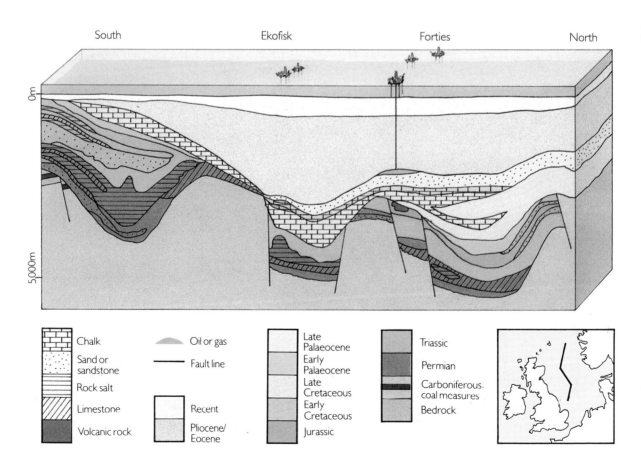

In the Cretaceous period that followed, about 120 million years ago, the North Sea was 40 degrees north of the equator, and once again isolated except for an opening to the north. Then water levels gradually rose until almost the whole of northwestern Europe was under a chalk sea. The only exceptions were the highest parts of Scandinavia, Scotland and Wales.

For some unknown reason, the chalk stopped forming – only one of many profound changes during a period that also saw the extinction of the dinosaurs. At the same time, the land masses of Europe began to appear in recognizable form. The North Sea ebbed and flooded alternately, connecting Britain with Europe during the Palaeocene period (58 million years ago), separating them in the Eocene period (45 million years ago), and then rejoining them in the Miocene period (14 million years ago).

Throughout this time, the North Sea maintained its drift to its current position. Then, about two million years ago, there came a succession of ice ages. In the most recent of these, which was at its peak 20,000 years ago, ice filled the whole of the sea north of the Wash to Jutland; between Scotland and Scandinavia, it was up to 2 kilometres thick, and it gouged deep valleys in the seabed, such as the 300-metre Devil's Hole, which lies off the coast of Scotland. Blocked by the ice, the rivers of northern Europe flowed along what is now the English Channel, carving a gully now known as the Hurd Deep to the north of the Channel Islands, and eventually meeting the sea at a coastline at least 100 metres lower than it is today, somewhere to the west of Cornwall and Brittany.

The ice had retreated from the North Sea by 13,000 years ago. Plough marks of icebergs beached during this final phase can still be traced off the coast of Norway and to the north of Scotland. As the seas returned, they pushed ahead of them the glacial relics of shingle and sand, providing much of the material for present-day coasts. Even today the fine details of the shores of the North Sea are still being created with every turn of the tide.

Ice wall
This glacier, at Briksdalbreen in northern Norway, gives an impression of the North Sea 20,000 years ago. Then the sea was filled with a mass of ice, plunging into the Atlantic at its northern margin.

ROCKY COASTLINES

Rocky landscapes fringe the North Sea, the only major gap being between Denmark and northeastern France. This exists because the mass of land centred on the Netherlands is slowly sinking, while the rocks all around have been pushed up – indeed, some are still rising.

The age of the rocks along the North Sea's coasts also reflects the vertical movements of the Earth's crust. Such cliffs that exist in the central area consist either of glacial deposits or of barely consolidated sand. Beyond this, in northern Denmark, on the Lincolnshire and Yorkshire coast of England, and both north and south of the eastern Channel, are spectacular white chalk cliffs, formed more than 65 million years ago. Further along the shores are the older Jurassic sandstones, rich in marine fossils, of the Seine basin, Dorset and Yorkshire. Carboniferous rocks, including coal deposits, meet the North Sea in England's Northumberland.

The oldest and most diverse rocks of all – granites, volcanic rock and modified sand and mud stones, at least 350 million years old – are at three of the North Sea's far-flung corners, in Norway, Scotland, and on both sides of the western Channel. As a result, a spectacular, if broken, rim frames the sea.

Variegated cliffs
Layers of sand and clay were deposited by the sea and then upended by titanic forces, creating the distinctive cliffs of the UK's Isle of Wight (above).

Where ice once flowed
These steely rocks on the Norwegian coast (below) have been ground into shape by the glacier that flowed over them during the ice age.

Stark contrast
The abrupt transition from red to white chalk in these cliffs on the East Anglian coast (below) marks a change from the deposition of minerals close to land, where sediments carried by the rivers gave the rock its red colour, to deposition in deeper, purer water further out to sea.

Where the wind blows
These precipitous cliffs (below) mark the northwestern edge of the North Sea, meeting the full force of the Atlantic storms at Hevdadale Head, in Shetland. They lie at the foot of Shetland's highest peak, the 500-metre Ronas Hill, and are composed of ancient volcanic rock.

Rock of ages
Chalk cliffs, such as these at Møns Klint in Denmark (above), are the remains of microscopic plants deposited in the sea over millions of years. Each 3 metres of chalk represents some 100,000 years of deposition.

A softer shore
Barely more than compacted sand, these geologically young cliffs (right) on the island of Sylt, in the German Wadden Sea, are rapidly eroding, exposing a brown layer of compressed peat beneath them, near the water's edge.

THE LOWLANDS

Where the rivers break through the cliffs to the sea, or the sea heaps up sediments against the land, or where the land is slowly subsiding, a coastal lowland forms. This landscape, with its unbroken horizon, subtle colours and shapes that change with the tide, is every bit as remarkable as that of the rocky coastline. The rocky shores actually help to create the lowlands – the actions of the waves tear the rocks from the cliffs and wear them down to boulders, pebbles, gravel, sand or mud. Coastal currents and waves sort the materials by size and push them along the shore to create spits and bars. Mud settles in sheltered areas and sand is blown on shore.

Unlike rocky coastlines, the physical shape and form of the lowlands is partly dependent on plant life. On sandy shores the leaves of the pioneering marram grass are toughened to resist the damaging effects of the abrasive sand, the heat and the aridity. The ever-enlarging clumps trap sand around them, and the dunes themselves grow up with the plants.

Similarly, the salt marshes are actually created by the marsh plants, which trap a thin layer of silt with each tide, and gradually raise the level of the flats out of the water. In this way, in suitably sheltered areas, both sand dunes and salt marshes slowly extend into the sea.

Land from the sea
Sand is blown up the beach (right) and settles around marram grass (*Ammophila arenaria*), which grows up with the dune. The grass roots stop the sand being blown inland.

The estuary mouth
As the river slows at the estuary mouth (below) it drops the sediments suspended in the water and they become trapped by salt marsh. The estuary gradually extends until a balance is reached between land creation and erosion by the sea.

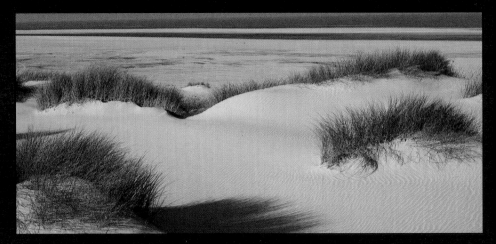

THE PRESENT AND THE PAST

Hard rock
These black boulders (left) are of basalt, a hard volcanic rock that is slow to wear down. Other less-resistant rock is reduced to sand by the waves.

Old duneland
As new sand dunes stride out to sea beyond the horizon (right), the older dunes lose their supply of fresh sand. They become covered in vegetation and pools of fresh water gather in the depressions.

The magic slate
The sea returns twice a day to wipe clean the record of events left on the surface of the sand flats (below).

The hinterland
In the sheltered conditions of the lowlands, vegetation flourishes along the waterline, either as salt marsh or, in less saline areas, as reed beds. In autumn the reed beds turn from green to gold.

THE SEA

The Romans regarded the North Sea as the roughest in the known world, a sea where storms and squalls blew up without warning, and one so huge that it was held to be uttermost, stretching to the ends of the Earth.

Two thousand years later we know of some of its more friendly characteristics. Itself warmed by water from the Atlantic, in winter it acts as a gigantic storage heater, keeping the northwest of Europe warm; in summer it reverses its role, keeping the land adjacent to it cool.

Nevertheless the North Sea does have a rough side to its nature. Hurricane-force winds blow up regularly in the winter months in the northern waters, accompanied by waves over 25 metres high. Tidal currents can also be fierce, flowing at more than 9 kilometres per hour off the Cherbourg peninsula and Portland Bill, and only a little slower at the mouth of the Elbe and the Pentland Firth. Large tides, backed up by a storm surge, can cause extensive flooding inland.

Around 7,000 years ago, an earthquake and huge landslip off the coast of Norway created a tsunami (tidal wave) that struck the east coast of Scotland, penetrating 4 kilometres inland and overwhelming the settlements there. If such an event were repeated today, the effect would be devastating.

Coast under pressure
The enormous power of the waves gradually breaks down even hard rock (above) as air is compressed within its chinks and crevices with explosive force.

Placid sea
A dead calm – a rare event in the North Sea. There are only between three and six days a year in the North Sea when the wind speed is 2 km/h or less.

THE PRESENT AND THE PAST 27

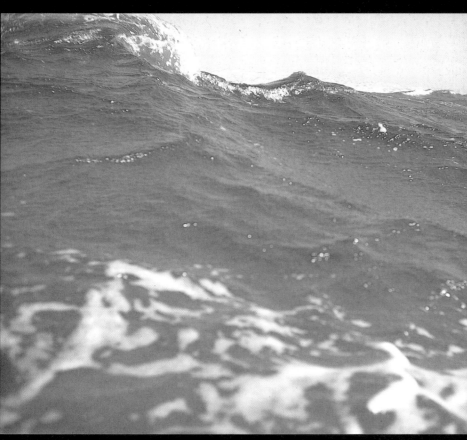

An ice blue wave
Of the two types of water in the North Sea, the oceanic water entering from the north (left) is clear, especially in winter, compared with coastal water, which is laden with sediments.

Deceptive tranquillity
These pebbles (above), brought down from the Scottish mountains, shift in rough weather, preventing life settling upon them.

The tidal race
Tidal streams (left) reach their maximum speed in areas where the water is forced through a narrow gap or around a headland. They can be strong even if the tidal range is small. Generally they flow at under 2 km/h in the north, 2-5 km/h in the south and 4-8 km/h in the English Channel.

Full gale
Storm-force winds whip up the waves in the North Sea (above). The size of the waves depends on the fetch of the water – the distance over open water that the wind has blown. Roughly once every 50 years, waves reach 30 metres high in the northern sea, and 10 metres high in the Strait of Dover.

THE TEEMING SEA

THE NORTH SEA supports a huge variety of life. At least 30 species of whale and dolphin, 6 species of seal and the otter have been recorded in its waters. There are over 70 species of seabird breeding around its fringes, and many more spend the winter there. Within its depths well over 170 species of fish have been found. There is a vast number of smaller animals, the more familiar of which include sponges, anemones, worms, shellfish, crabs, lobsters, starfish and sea-urchins. A column of water with a surface area of just one square metre may house between 750 and 2,700 creatures of different kinds. In the same area there may be millions more within the sediments on the sea floor.

These species have evolved over millions of years, in ways that allow them to live in different regions of the sea, different environments such as muddy, sandy or rocky seabeds, or in the lower, middle or upper reaches of the water.

The zones of the North Sea

It is the extraordinarily diverse physical nature of the North Sea that has attracted such a wide variety of species to its waters. In the north, the protruding tongue of food-rich Atlantic water entices whales, and supports huge colonies of guillemots, puffins, razorbills, gannets, and gulls. In the south, the warm waters of the western English Channel draw species ranging from coral to pilchard. Between these two extremes, dense shoals of silver-indigo herring ply their way through the surface waters, pursued by seals and dolphins, themselves ever ready to take flight from the menace of the killer whales.

Life in the North Sea
The North Sea is not an amorphous mass of water, but is separated into zones, each of which suits a particular set of creatures. The different types of seabed and varied coastline provide habitats for a huge range of plants and animals.

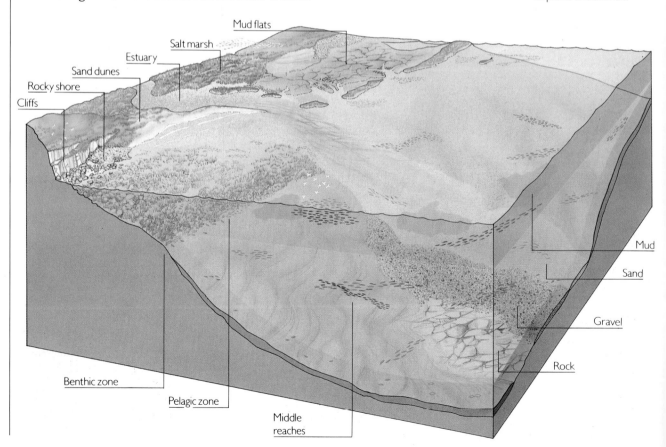

Sharks such as the tope and spurdog prowl the middle reaches. Cod and haddock, ling and pollack lurk deep below, scanning the bottom for tasty morsels. Norway lobsters and burrowing crabs tunnel through the mud. Rag worms and cat worms, razor shells and cockles thrive in the sand. Sponges, anemones and fan worms attach themselves securely to the rocks. Starfish, brittle-stars, sea-urchins and crabs clamber over the tumbled remains of generations of whelks, scallops and other shellfish, disturbing myriad smaller species in the process. On the bottom, flat fish and skates, angler fish and eels forage for food, along with rarer species such as sturgeon and electric ray.

In shallower waters, where light can penetrate, kelp forests sway in the currents of rocky coastlines, and eel grass binds together muddy and sandy sediments. Even on these less hospitable fringes of the sea, animals such as barnacles, mussels and crabs make a living – their reward is a table that the tide resets twice a day.

THE HEART OF THE FOOD WEB

Such a diversity of animals and plants is amazing. But more remarkable still is that this extraordinary range of species, and all their various interrelationships, depends almost completely on minute, single-celled plants suspended in the top layer of water. Called phytoplankton, they are not just the beginning of a conventional, linear food chain in which big eats small. Instead, they form the very heart of a three-dimensional food web.

Phytoplankton – the organic alchemists

Following the rhythm of the seasons, phytoplankton trap the energy from the sun and use it to turn mineral nutrients and carbon dioxide into life. On average every year a kilogram of carbon is captured beneath every 10 to 15 square metres of sea surface and turned into organic matter. Bacteria and other minute organisms (for our purposes, honorary members of the phytoplankton) may add significantly to this amount. In this way the phytoplankton directly or indirectly provide the food for most other creatures living in the North Sea.

Three main groups of phytoplankton are common: the diatoms, the dinoflagellates and the microflagellates. The diatoms are named after the two separate parts of their silica shell, one of which fits inside the other like an old-fashioned pill box. Diatom species may range in size from a few thousandths of a millimetre to, exceptionally, over a millimetre. The second main group is the dinoflagellates. These have small, whip-like threads, called flagella, with which they propel themselves through the water. These 'plants' may also have a primitive eye: in this microscopic world, the distinction between animal and vegetable can be a hazy one. Individuals of the third group, the microflagellates are, as the name suggests, very much smaller than those of the other two. This group is made up of several kinds of phytoplankton, differing in form and habit. Some are soft while others have a toughened outer wall. Some use photosynthesis to create their own food; others have become carnivores, feeding on fellow members of the plankton. Within the groups, the exact identification of phytoplankton species can be very complex: unless it is essential, scientists tend to distinguish them by a code based on morphological characteristics, thus LRGT1, LRGT2 etc, where LRGT stands for 'little round green thing'.

Phytoplankton grow and multiply rapidly. Their numbers double at a rate anywhere from a few hours for the smallest species, to a few days for the largest. Given their importance within the food web this is just as well.

Zooplankton grazers

The food web is actually made up of a number of small chains. The first vital link is made by the zooplankton, communities of tiny animals that drift through

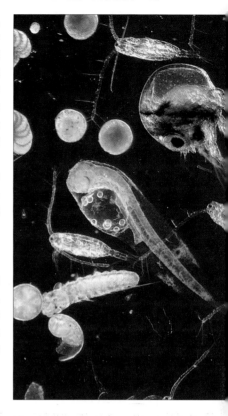

Mixed catch
Virtually all life in the North Sea depends on plankton, mostly microscopic animals and plants that drift or float along in the water (above). Minute green microflagellates and round diatoms (both shown above), together with dinoflagellates (below), form the main types of plant or phytoplankton. Among the animal or zooplankton in this sample are copepods, which are small crustaceans, and the larvae of marine worms, crabs and fish.

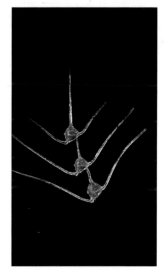

the water with the phytoplankton. Many of these are permanent members of the plankton; some are larval forms of creatures – fish, worms, crustaceans, jellyfish – that will later leave, to spend their lives in the lower reaches or on the seabed.

Of the zooplankton, one group in particular grazes upon the phytoplankton: the copepods. These small, mainly herbivorous crustaceans are by far the most abundant of the different kinds of zooplankton: in a cubic metre of water there may be many millions of adults and young. Indeed, copepods are the most abundant multicellular life-form in the world, far more numerous than even insects. These creatures range in size from that of a pin head to that of a grain of rice. They filter the phytoplankton from the water, using intermeshing bristles on their legs, and may actually distinguish between individual species of plankton, choosing only those whose walls are without armour plating.

Carnivorous zooplankton

The copepods feed mainly on the phytoplankton; just about everything else in the zooplankton then feeds on them. In contrast to the copepods, the rest of the zooplankton take a huge variety of forms, capturing their food in ingenious ways.

Of the jellyfish, for example, the species *Aurelia* and *Cyanea* look much like the traditional image of a jellyfish, although their tentacles act rather like flypaper, capturing the plankton that brush against them. But the species *Rhizostoma* has thousands of minute mouths suspended beneath it, resembling a cross between a football and a mushroom, drawing in water and copepods as if into a sponge.

Comb jellies, including species such as the sea gooseberry, fish for their food. They trail long lines down from their bodies, each equipped with rows of lasso cells that trap copepods and anything else unfortunate enough to swim against them. From time to time, the comb jelly draws the lines up to its mouth, and devours any captives it has made.

Eat and be eaten
Copepods such as *Acartia* feed on plant plankton, and themselves form the staple diet of nearly every other member of the plankton.

Giant plankton
Jellyfish such as the lion's mane (*Cyanea capillata*) are the largest creatures of the plankton, growing up to 50 centimetres in diameter.

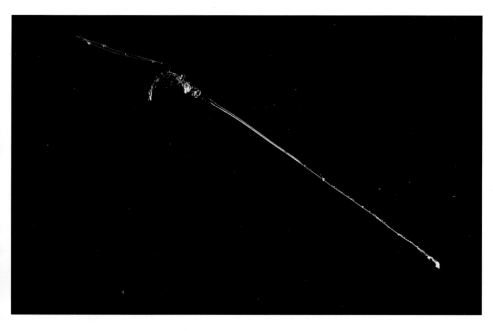

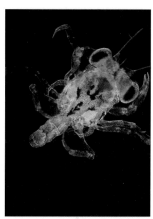

Young and old
Crabs go through an extraordinary metamorphosis. The crab larva has a horn (left) that helps it to stay in the surface waters, with its food. As the larva grows (below), the horn disappears and the crab swims actively for its prey, which it seizes in its pincers. Later it will tuck its tail underneath its body, and settle on the bottom for its adult life.

The arrow worms form a remarkably old and highly successful group of voracious predators. They were present in the Cambrian era, one of the earliest periods in which fossils are known. Since then they have seen other groups come and go, and have probably tasted them all. But during their 500 million years of existence they have changed very little. Their elongated form, like an arrow, allows them to dart forward and grasp prey with a pair of viciously hooked claws. They play a major predatory role in the plankton community, and inflict considerable losses even on larval fish such as the herring.

Other forms of zooplankton range from protozoa to swimming worms and molluscs, the latter including species that are perfectly formed small snails, except that their 'foot' is expanded into two wings, with which they flap their way through the water. Many of these species seem bizarre and exotic, like creatures from another world; indeed, they are found nowhere else but the open sea, and are highly adapted to their peculiar, watery medium.

Just passing through

Perhaps the most remarkable feature of all in this strange world is that some species spend only part of their lives there, sharing the larder and the predatory risks of the zooplankton shoals. It is these creatures that form the vital link between zooplankton and the whole of the North Sea ecosystem.

Some of the plankton go through such extraordinary metamorphoses that it is hard to relate the adult form to the young. In the case of crabs, the puzzle of where the young came from was solved only in the 19th century. Until then, all that was known was that crabs laid eggs and that small, perfectly formed crablets would suddenly appear some time later. The stages in between were a mystery. But now we know that the egg hatches into a tiny larva, superficially resembling an underwater mite. This then goes through two, four or five stages, growing a little with each moult, looking at first glance like a shrimp, but with a 'unicorn's horn' arising from its head. The horn of the porcelain crab is swung forward and elongated as if ready for some planktonic jousting tournament. These long spines may help to prevent the juveniles sinking in the water, but they may also make the creatures difficult to eat. In its final stage as part of the plankton, the organism looks a little more like a crab, but it has a tail stretched out behind it. Prawns, shrimps and lobsters are among many other crustaceans that go through similar – if less dramatic – metamorphoses.

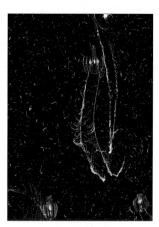

Ready and waiting
Sea gooseberries (*Pleurobranchia*), a species of comb jelly, will ensnare any creature that brushes against their waiting tentacles.

THE DANCE OF DEATH

If we are to understand the nature of the life in the North Sea, we must do more than sit back and marvel at its extraordinary diversity. We need to look closely at the ecology of the sea, to learn how its various creatures interact among themselves, as well as with other species. Until we know much more about how the whole web of life in the North Sea works, we cannot tell what impact our own actions will have on it.

In the long term, leaving aside the human influence, the various species have more or less steady populations: the creatures that are plentiful or rare today were equally abundant or scarce a century ago. Two factors help to control their numbers: 'chaotic' and 'regulatory' death.

The effects of chance

Chaotic death is death at the throw of the dice; it is unexpected and unpredictable, caused, for instance, by the vagaries of the weather. One year a species may be devastated, the next it may escape unscathed. This type of death can strike any species – no matter how large or small its numbers.

The controlling factor

Regulatory death is different: the more individuals of a species there are, the greater the number that die. Only a fixed number of barnacles can fit onto the boulders that make up a rocky coastline. The rest – whether there are hundreds or thousands of them – all perish.

Regulatory death is often brought about by the actions of other species. Lethal diseases and parasites, for example,, spread most easily in densely crowded populations. On some rocky shores predatory starfish prowl, feeding exclusively on the creature that happens to be the most abundant at the time. In this way, the numbers of mussels and other shellfish on the shore are kept within a certain limit; beyond that, they attract the attention of the starfish, which bring their numbers down. Similarly, grazing sea-urchins can control the quantity of kelp weed on a rocky coast. Starfish and sea-urchins are examples of 'keystone' species, which may control the population of whole communities, even though they themselves may not be particularly common.

A tangled web

These two forces, chaotic and regulatory death, act in opposition. Chaotic death disrupts the 'normal' levels of populations, while regulatory death returns them to their typical levels, be they common or rare. The problem for anyone wishing to apply this knowledge to protect the North Sea is that in practice it is very difficult to

Danish mud flats in winter
In a hard winter, the mud flats may freeze (top right), destroying whole groups of cockles and other creatures that live there.

The battleground
The diversity of species found on a rocky seabed (right) is the result of a constant battle for life and death, in which none can gain the upper hand.

Complex fate
Dog whelks (*Nucella lapillus*) prey upon barnacles (*Semibalanus balanoides*, above). Predation, overcrowding, even a violent storm, all affect the fate of the barnacles.

The need for space
Some fish, such as the corkwing wrasse (*Cranilabrus melops*, left), are territorial, but the sea has room for only a fixed number of territories; beyond that, all contenders for the space will die.

disentangle the two kinds of event. Because every single creature – predator, prey and parasite alike – is affected by both chaotic and regulatory death, the system as a whole is highly complex, and constantly changing.

For a whole series of reasons it has proved virtually impossible to work out exactly what – predator, parasite or disease – controls the populations of individual species, particularly those that live in the sea. Without knowing this, we cannot predict the likely effect of discharging toxic waste or of fishing, or of any action that we take, on the North Sea. Until we can make that prediction, we must learn to be far more cautious about what we do: then, the life in the North Sea might stand a better chance of survival.

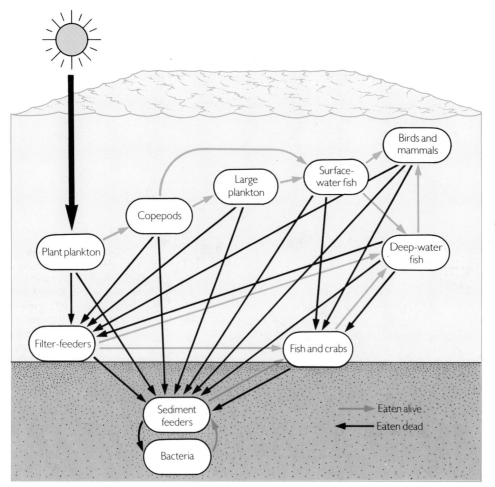

The North Sea food web
The sun is the source of all life in the North Sea. Plant plankton harness its energy and use it to create carbohydrates. They are fed on by animal plankton, primarily the copepods, and these in turn are fed upon by larger plankton and fish. Debris and bodies fall to the seabed, providing the animals that live there with food. Some creatures such as crabs and fish will feed directly on large food items; others, such as anemones, shellfish and fan worms, rely on filtering minute particles from the water. The filter-feeders themselves are devoured by large fish. The waste produced by all these animals provides the nutrients for the next generation of plants.

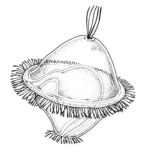

A primitive form
Worms and shellfish look totally different from one another when they are adults, but many have the same basic form in their earliest larval stage (above).

Crustacean larvae, even those of the crabs, are visibly crustaceans. But many other larvae that live among the plankton are so different from their adult forms that it is hard to tell even to which group they belong. Most striking of these are the 'trochophore' larvae of some primitive worms and molluscs. Each one is shaped like a top with a fringe of short flagella around its middle and is crowned by a tuft of sensory hairs. A mouth opens on the outer rim; the gut extends right through the body, and re-emerges at its base. The larvae of all other molluscs and worms, although more complex, are clearly modifications of this basic form.

The ghost of a creature
The larva of the crawfish (*Palinurus vulgaris*, right) is ideally suited to life among plankton, its transparent form making it barely visible to either predators or prey.

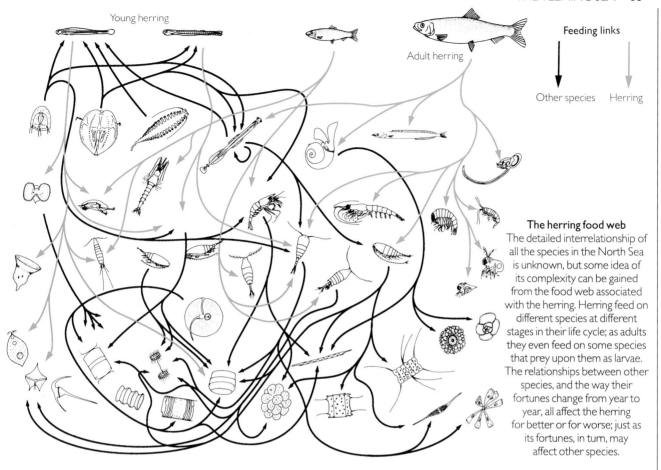

The herring food web
The detailed interrelationship of all the species in the North Sea is unknown, but some idea of its complexity can be gained from the food web associated with the herring. Herring feed on different species at different stages in their life cycle; as adults they even feed on some species that prey upon them as larvae. The relationships between other species, and the way their fortunes change from year to year, all affect the herring for better or for worse; just as its fortunes, in turn, may affect other species.

They include the rag worms and lugworms, limpets, dog whelks and oyster and most other worms and molluscs found in the North Sea.

The larvae of echinoderms – the family of starfish and sea-urchins – show yet another extraordinary twist. In this case, a delicate branching larva, as clear as glass, starts to develop, complete with senses, gut, and swimming organs. It goes about its business, hunting and being hunted in the normal way. But then a growth starts to form in its side, and it is this that is destined to become the adult form. In time, the growth also acquires its own mouth, senses and swimming bands; growing at an ever faster rate, it soon absorbs its original host.

Many fish, too, begin their lives as plankton. Flat fish, members of the cod family, and the herring family, along with sandeels and many others, all pass through a larval stage among the plankton, where they pursue the copepods and are eaten by other, larger zooplankton, such as the arrow worms. The fish time their reproduction to allow their juveniles to enter the plankton when there is plenty of food available, so that they pass through this stage, when they are extremely vulnerable to predation, as rapidly as possible.

A species such as the cod starts its life by feeding briefly on diatoms and dinoflagellates, before moving on to larval copepods and molluscs, and other zooplankton of the same size. As the cod larva grows, it changes to a diet of adult copepods, crab larvae and other zooplankton. The only requirement is that the prey should be slow enough to capture and small enough to swallow.

At the end of their sojourn in the plankton zone, all of these creatures graduate to their adult forms, and then a massive migration takes place. Shore crabs scurry off to shelter on the rocky shore. Cod swim down to deeper waters, while rag worms and flat fish, prawns and shrimps, starfish and sea-urchins drop to the seabed. Established in their new surroundings, these species are ready once again to pick up their lethal game of eat and be eaten.

THE OPEN SEA

Other fish may leave the world of the plankton, but species such as herring remain behind, feeding on the community to which they once belonged. Herring are attracted by dense patches of large copepods, although they will also feed on other plankton that swim with them. Species such as the arrow worms, once savage predators of the herring in its early life, now become just another snack.

By changing its role from prey to predator, the herring arrives at the top of the plankton food web. But it now finds itself at the start of a much larger chain, which ends with a host of larger predators that ply the open sea.

Creatures in demand
Slender-bodied sandeels (inset) swim in huge shoals in the shallow waters of the sea. There are five species in the North Sea, each of which is food for other fish, birds, seals, dolphins and porpoises.

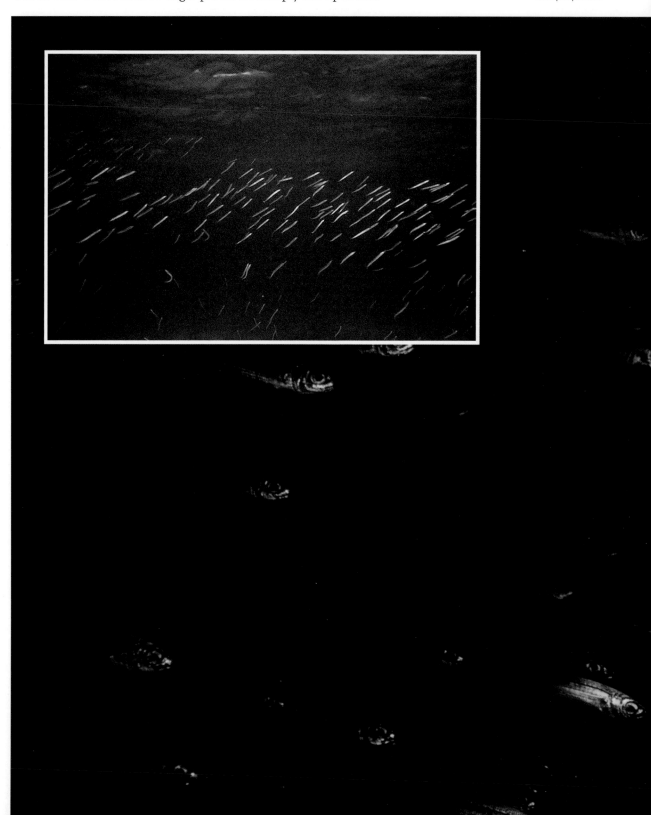

Cold-blooded killers

The fish that, as adults, turn their backs upon the plankton, feed instead on the sandeel and herring, and on the herring's smaller relative, the sprat. The shoals of sandeel in shallow waters represent a highly desirable meal for anything large enough to swallow them, and they spend their lives hiding in the sand from predators unless they are feeding or mating. Two of the major species to feed on the herring are the cod and the mackerel, once its companions in the world of the plankton, and the herring also forms a significant part of the diet of other members of the cod family, including pollack and saithe.

Safety in numbers
Herring and mackerel (below) are two of the commonest fish in the surface waters. Mackerel will feed on the herring if they are small; otherwise the two species swim together in shoals for protection against predators, which become confused by the mass of darting fish, and cannot focus on any individual.

SEALS

Two species of seal breed in the North Sea, the grey seal (*Halichoerus grypus*), and the harbour or common seal (*Phoca vitulina*). The grey is the larger, the male growing up to 2.45 metres from nose to tip of the tail flipper, and weighing 310 kilograms; while a male harbour seal can reach 1.85 metres and weigh 105 kilograms. Males of both species may live for about 20 years; females for more than 30 years.

Seals spend a great deal of time in the water, which makes them difficult to count, but there are thought to be approximately 90,000 grey seals in the North Sea; harbour seal numbers were around 40,000 before the epidemic of 1988, when 18,000 died. The grey seal is most abundant in the northwest North Sea, and is typically found around weather-beaten rocky coasts and caves, while the harbour seal occurs throughout the North Sea, especially around sand bars, mud flats and estuaries. A few greys live in the western English Channel, where the harbour seal is only an occasional visitor.

One feature that helps to distinguish between the two species is the profile: the line from the grey's forehead to its nose is virtually straight, while the harbour seal has a longer muzzle, with a dip between brow and nose.

Salmon feast
A grey seal devours a salmon (above). Seals have a catholic diet of various members of the cod and salmon family, herring, flat fish, sandeels, molluscs, crustaceans, sea-urchins, and even octopus. Fishermen may blame seals when their catches are small, but there is no evidence that seals have a drastic effect on fish populations.

Graceful transformation
A harbour seal in its element (left). The seal uses its hind flippers to propel itself along under water, frequently swimming upside down to peer at the bottom for food. When seals are on land, they move by hitching themselves forwards using their front flippers.

Mottled harbour seals
Seals have short coarse hair, but rely on their fat layer for insulation against the cold.

Resting on the shore
A large group of grey seals hauled out on a rocky beach. The seals moult in summer, the old fur peeling away in a layer.

Grey seal pup
The grey seal pup is born with a white coat, which will later be replaced by mottled grey fur. The mother seal's milk is so rich that the pup grows from 14 to 45 kilograms in the space of three weeks. Then the pup must fend for itself as the female mates again.

Superimposed on this structure is a hierarchy of ever larger fish feeding on other fish. Abundant small sharks such as the tope (which grows up to 1.5 metres) and spurdog (up to 1.2 metres) swim in shoals over the full depth of the sea, feeding on herrings and sandeels in the surface water, cod and its relatives in the depths. These sharks in turn fall prey to even larger sharks. But while the fish-eating fish form a major part of the food web, they do not have the feast all to themselves.

Air-breathing fish

As far as fish are concerned, seabirds such as the auks, northern homologues of the penguins, are air-breathing fish that swim in packs, pursuing and scattering the shoals, matching every twist and turn of the individuals desperately trying to avoid their clutches.

Auks are the most plentiful offshore seabirds, and of these the guillemot is more plentiful than the rest put together, with a population of more than two million. During the breeding season in May and June, guillemots gather food for their young from coastal waters, travelling 10, 20 or 30 kilometres to favourable

White death
Two gannets (*Sula bassana*) surface after a successful dive (above), one swallowing its prey, the other about to do so. The northwestern North Sea is an important breeding ground for gannets: each year, as many as 88,000 birds will congregate there, attracted by the larger fish – herring and cod fish – to be found in its waters.

Nursery on a cliff edge
A crowded ledge on a steep cliff face (right) provides a safe breeding ground, well away from predators, for a colony of guillemots (*Uria aalge*). Each pair of birds lays a single egg, not in a nest but directly on bare rock; the egg's conical shape reduces the chance of it being knocked off the ledge and plunging into the sea.

sites. After about three weeks the chicks are large enough for all the birds to leave the cliffs and move out to wherever the greatest concentration of fish may be, to lead a life in which swimming is more important than flying. During July and August the adults dispense with flying altogether, simultaneously moulting all of their flight feathers. Sandeels by now are disappearing into the sands, and the guillemots take an increasing proportion of herring and sprat instead, along with any small cod that come within reach. By the winter, many guillemots are in the southern North Sea, feeding on sprats. Then, as spring returns, the population shifts back to its breeding sites, although before the chicks have hatched the adults may range over a radius of 100 kilometres to devour the re-emerging sandeels.

The feeding habits of other species of auk, including puffins and razorbills, show variations in detail rather than substance. Most dramatic of all the birds is the huge gannet, which crashes through the sea's ceiling to snatch an individual from a shoal – one moment there, the next gone without trace. The gannet is the only bird that will regularly take fully grown herring. At the other extreme, delicate terns concentrate on diminutive sandeels.

Selkies, cetaceans and leviathans

Occasionally an auk swimming on the surface may suddenly disappear from view – not because it is diving for food, but because it has itself been snatched, often by a grey seal. There are two species of seal that breed in the North Sea: the larger grey seal is usually found in northwestern waters, and the harbour or common seal in the east and south. Seals enjoy a catholic diet of fish, molluscs, sea-urchins, even shrimps. Their senses are remarkable. They can hunt successfully in the dark and in opaque water, and there are records of completely blind grey seals not only surviving, but living a normal life including rearing a pup and providing it with food. The secret of their success in hunting lies partly in their wet nose, which allows them to pick up scents even while diving, when their nostrils are closed. Long whiskers help, too, sensing a change in the currents and guiding the seals to their target in the final stages of the chase.

The amphibious seals return to land in order to breed, but dolphins and whales live exclusively in open water. Among the more common species are white-beaked dolphins, found in the northern North Sea; common and bottlenose dolphins, in the English Channel; and harbour porpoise in coastal waters. All have varied diets, except Risso's dolphin, a resident of the English Channel, which is believed to feed exclusively on cuttlefish. Perhaps the most fearsome to all is the killer whale or orca, which is actually a large dolphin. The males grow to 9 metres

Breaking the waves
The common dolphin (*Delphinus delphis*) is most usually found in the English Channel. The markings on the body, clearly visible when the dolphin leaps from the water, are most distinct during the breeding season.

Top predator
The male killer whale or orca (*Orcinus orca*) is the largest of all dolphins, and has an exceptionally tall fin. The orca has been seen to play with its prey like a cat with a mouse, but it has never been known to harm humans.

in length, the females to two-thirds that size, and the dorsal fin of the males may be as tall as 1.8 metres. The orca is a worldwide species that is known to feed on large fish, seals and dolphins and even, when in packs, to hunt whales. But just what it feeds on in the North Sea is far from clear.

Last in the range of cetaceans that visit the North Sea are the leviathans – the baleen whales. The smallest of these in the North Sea, the minke whale, is still a giant, reaching 9 metres. This is yet another species that concentrates on the herring for food, although it will eat other creatures. Only about 30 years ago, the minke whale was common in the northern North Sea, coming as far south as the waters off the Wash and Wadden Sea.

The second largest whale in the world was once abundant enough to justify the name of common rorqual, but its numbers – like those of the minke – have been severely depleted. Individuals were once described as 'not so rare' in inshore waters, turning up in the North Sea all the way down to the Strait of Dover. This 24-metre giant feeds on herring, hence its second name 'herring hog', and directly on plankton.

The plankton-feeding baleen whales complete the food web of the open waters of the North Sea. But that is not the end of the marine chain: from this teeming life in the surface waters a constant rain of organic material – faeces, dead and dying plankton and fragmented bodies of fish – falls to the seabed.

LIFE IN THE BENTHIC ZONE

For the creatures that live on the sea floor, in the region known as the benthic zone, the structure of the seabed itself – whether of mud, sand, gravel, pebble, rock, or a mixture of these – presents different challenges. Rocks provide a firm point of attachment for some animals, where they wait for food to come to them rather than seeking it out; but because they cannot swim away, the creatures themselves are at grave risk from predators. Sand provides no such anchorage, but large numbers of small species can live between its grains. Whatever its composition, if the seabed lies at a depth greater than about 30 metres, there will not be sufficient light to allow plants to create food by photosynthesis, and the benthic community will be dependent on the food raining down from above.

Making a home in the mud

It is rare for large food items to reach the seabed intact. When organic debris falls in really deep water with a muddy bottom, like that of the Norwegian Trench, it will scarcely have hit the seabed before a host of animals, including Norway lobster, crabs, brittle-stars and gruesome hagfish and ratfish, crowd around it, frantically tearing and devouring it. Tiny particles drift away in the current, eventually to fall again to the bottom where they will be fought over by a collection of minute crustaceans, even smaller thread-like worms and single-celled animals, all of which live in intimate association with the sediment through which they suck, lick and nibble their way from one generation to the next.

Oils, fats and minerals ooze down into an even deeper, evil-smelling black zone, where there is no oxygen and the bacteria that live there survive by breathing sulphur and other alien compounds. This is the furthest extreme of the marine chain. Beyond this, no form of life can possibly survive.

Above, in the more hospitable sediment where there is oxygen, various bristle worms search out and eat the small crustaceans, while deep-water flat fish such as the witch and megrim, along with long-nosed skate, will take both worms and crustaceans. Large cod and haddock patrol these depths, swimming just above the bottom, devouring crabs, worms, small flat fish and virtually anything else that is remotely edible. In this way, the benthic food web, built on the debris falling from the surface layer, in turn feeds creatures living in the higher reaches.

Giant filter-feeder
Unlike other sharks, the huge basking shark (*Cetorhinus maximus*, left) has only vestigial teeth. It feeds entirely on plankton, gathered by taking in huge mouthfuls of water; the plankton are filtered from the water in the enormous gills and then consumed in one enormous gulp. Basking sharks fast through the winter months, when plankton are scarce.

Peering from shelter
Norway lobster (*Nephrops norvegicus*) digs burrows in the mud in the sea floor – some may stretch back as much as a metre.

46 THE GREENPEACE BOOK OF THE NORTH SEA

Fans of colour
Peacock worms (*Sabella pavonina*, right) in search of food extend their tentacles from the protection of their tubular casing. A sticky liquid is sent down the length of each tentacle and returned along the opposite side, complete with food.

A subtle disguise
Depths of up to 35 metres provide suitable conditions for the squat lobster (*Galathea strigosa*, left). It keeps out of harm's way under stones and rocks, emerging with the need to find food. The claws and feet are camouflaged to disguise its shape.

Filter-feeders on the seabed
Queen's scallop (*Chlamys opercularis*), sea anemones (*Protanthea simplex*) and dense, branching colonies of their smaller relatives (below), all feed by filtering minute particles of organic matter from the water.

FISH

There are more than 170 species of fish in the North Sea, of all shapes and sizes, and all incredibly different from one another. Lampreys and hagfish both have a rudimentary skeleton of cartilage, but they do not have jaws; lampreys attach themselves to a fish with their sucker-like mouth and then suck the blood from it, while hagfish are voracious scavengers.

At the other extreme, the bony fish have evolved a range of aids to survival. The swim bladder, a bag of air inside them, makes the fish buoyant, enabling them to maintain their depth effortlessly; and they have a fine degree of control over movement, able to hang in the water, or change direction with startling speed.

Many of these fish, including herring and mackerel, have a good sense of hearing, and generate sound by a variety of methods. Flat fish can change colour to match their surroundings; others, such as wrasse, have fixed but stunning colours, used in territorial display.

The salmon family has mastered the art of moving from salt water to fresh. The eel has gone one better, and can live temporarily on land, which allows it to migrate all the way from the Sargasso Sea, off the coast of Mexico, to lakes in the countries bordering the North Sea.

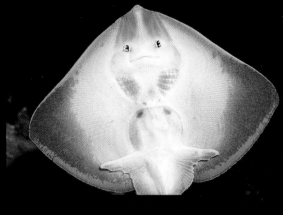

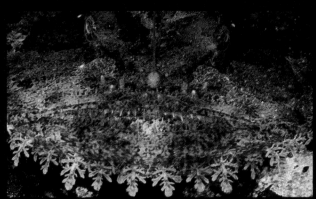

Flat shark
The underside of the common skate (*Raja batis*, top) reveals its mouth, nostrils (used for smell) and gill slits. The diamond shape is created by two enormously enlarged front fins.

The great deceiver
The angler fish (*Lophius piscatorius*, above) seizes its prey in its huge jaws. The peculiar fringe beneath the mouth breaks up the fish's shadow and disguises its presence on the seabed.

The caring parent
The weed of the rocky shore provides an ideal habitat for the lumpsucker (*Cyclopterus lumpus*, left), so-called because it has a sucker on its belly with which it attaches itself to the sea floor. The fish is also known as the sea hen, because the male tends the eggs until they hatch.

THE TEEMING SEA 49

Spot the fish
The long-spined sea scorpion (*Taurulus bubalis*, left) lives among the rocks and weeds below the low-tide mark, and has extraordinarily effective camouflage.

Wolf fish
The wolf fish (*Anarhichas lupus*, below) has powerful jaws and crushing teeth, with which it tackles whelks, scallops, mussels, sea-urchins and crabs.

Tools for survival
The sea lamprey (*Petromyzon marinus*, above) hangs onto rocks with its mouth, in order to maintain its position in fast-flowing currents. The holes on the flank are simple gills, which pass straight through to the throat. The unpigmented skin on the head is the lamprey's 'third eye', which produces hormones in phase with changes in the light and allows the fish to set its body clock.

King worm
A voracious predator that roams the seabed, the king rag worm (*Nereis virens,* above) grows up to 40 centimetres long. Each segment of its body has a pair of protrusions with bristles at the tips, which help the worm to propel itself along.

Not what they seem
Cuttlefish (*Sepia officinalis,* below) are not fish but intelligent molluscs. They change colour to blend in with their surroundings.

Second-hand value
The common whelk (*Buccinum undatum,* above right) is one of the largest shellfish living down to 100 metres. The empty shells often provide sanctuary for soft-bodied hermit crabs.

An unblinking guard
The tiny spheres within the rim of the great scallop (*Pecten maximus,* right) are actually eyes, enabling the scallop to detect predators approaching from all directions.

A settlement in sand and gravel

Sand, mixed with varying degrees of silt or gravel, forms the predominant seabed of the North Sea. On pure and silty sands and on gravel beds, the creatures known as filter-feeders find their ideal environment. Of these, the bivalve molluscs – represented here by species such as common, spiny and prickly cockles and Venus shells – are the most sophisticated.

Adult bivalves, a few centimetres in length, filter-feed on food particles only a few thousandths of a millimetre in size, a disparity at least as great as that between baleen whales and their zooplankton prey. The food the bivalve eats collects on the finely corrugated surface of its highly modified gills. Small particles, drawn through the curtain of the gills by water passing through them, are trapped within the grooves and are propelled on a 'conveyor belt' of mucus to one side of the gill by minute, beating cilia or hairs. From there, more cilia propel them along to the bivalve's mouth. Worthless, inorganic particles are typically too large to penetrate the grooves. Instead, these are propelled in the opposite direction by cilia and expelled outside the bivalve's body.

In soft sediments, bivalves may stay below the surface, and have syphons to carry water to and from them. In the depths, the filter-feeders make use of the organic debris; but in shallower water they may feed directly on the planktonic food web, and species such as cockles may then form huge beds on the sea floor.

Not all bivalves are filter-feeders. Tellins, for example, which live in clean sands, have an elongated syphon that they sweep over the surface of the sand, picking up food particles rather like a vacuum cleaner. The disadvantage of this is that flat fish will graze on the dense mass of waving syphons. Nor are bivalves the only inhabitants of the sand. As with the muddy seabed, a wide range of small creatures lives within the sediment, including an even wider range of crustaceans, predatory cat worms and rag worms.

Many species of flat fish, ray and skate spend most of their lives here on the sandy sea floor. A few show clear preferences for certain foods, which they unearth from the sediments. The lemon sole has a penchant for bristle worms,

A one-sided view
The flounder (*Platichthys flesus*) lies in wait for prey. Flounder, like other flat fish, start life among the plankton as normal fish. But then they change, and start swimming on one side. Eventually one eye migrates across the head to join the other – and the fish is now equipped to live on the seabed.

CRUSTACEANS

Crustaceans are the most primitive of the arthropods, a vast group of animals that also includes insects, spiders, scorpions, centipedes and millipedes. Crustaceans in the North Sea range in size from minute copepods, which are part of the plankton, to the decapods – shrimps, prawns, crabs and lobsters – that roam the sea floor.

All decapods have five pairs of legs for walking, as well as smaller appendages at the head, for feeding, and under the tail, for carrying eggs in the case of the female. The head and body of 'normal', shrimp-like, decapods is covered over by a single piece of shell, the carapace, with the multi-segmented tail extending behind. In adult crabs the tail is small, and tucked away underneath the body.

The shell, impregnated with calcium carbonate in larger crustaceans, gives valuable protection from predators. However, it does not allow room for growth. As a result, each creature has to shed its shell from time to time. The old shell becomes soft and the crustacean takes in up to 70 percent of its weight in water through its gut, causing the body to swell up, splitting and shedding the old shell in the process. The new shell, which has formed underneath, then hardens, and the water is excreted, leaving room for further, real, growth.

Climbing crab
The long thin legs of the spider crab (*Hyas araneus*, left) allow it to climb the rigid holdfasts of the kelp forest.

Defending her eggs
A female velvet swimming crab (*Macropipus puber*, below) rears up, displaying the fertilized eggs carried under her tail. The larvae will eventually be released into the world of the plankton.

The power of the claw
Antennae help the lobster (*Homarus gammarus*, top) to detect its prey – and exceptionally strong pincers enable it to break open the toughest of shells.

Food for free
The edible crab (*Cancer pagurus*, above) will seek out live food, but it is also a scavenger, taking dead and dying creatures that have fallen to the sea bottom.

THE TEEMING SEA

Hand to mouth
The pincers of the shore crab (*Carcinus maenus*, above) make short work of mussel shells; smaller feeding claws around the mouth then tear the food to pieces. The shore crab is abundant on rocky coasts.

See-through prawn
The transparent shell of the common prawn (*Leander serratus*, left) reveals the brain, above the eye, with nerves running horizontally from it, and the gut passing to the green digestive gland.

A soft co-operative
The soft coral known as dead man's fingers (*Alcyonium digitatum*, above) is made up of many small colonial sea anemones embedded in a flexible matrix that they secrete as they grow.

A snail without a shell
The sea lemon (*Archidoris pseudoargus*, right) is a sea slug that grows up to 7 centimetres long. The eyes are on soft horns on the head; the gills are clustered on its back.

and a small mouth with which to pick them up. The eagle ray, in the English Channel, feeds only on molluscs and shellfish, for which it competes with humans. But many species, although they are associated with different sediments, different water depths or different temperatures, share a mixed diet.

Gravel beds represent a harsh, abrasive environment at the opposite extreme to muddy sediments. Very few animals live within them, or even on their surface if they are washed by strong currents, as they are in some parts of the English Channel. Where conditions are calmer, brittle-stars may form dense beds on the gravel, from which they capture food suspended in the water. Despite the paucity of life there, gravel beds do play an important part in the ecology of the North Sea. Herring use them as their spawning grounds: the eggs fall through the water to the bottom and adhere to the small stones.

Rock bottom

In coastal waters, extensive areas of bedrock may be exposed. If it is chalk or another soft rock, it may be mined by piddocks (a kind of bivalve) and even boring sponges, which make their homes there, filtering food in the comparative security that it offers.

Where the seabed is uneven, it is a jumble of shells, rock, and an ooze of oils, minerals and clay. In deeper waters life is still dependent on the material falling from the planktonic zone above. Attached to any firm point is a mass of sponges filtering the water for food. Feathery anemones, soft corals such as dead man's fingers, and tube worms reaching from their twisted limestone casings, all capture food on outstretched tentacles. Barnacles lie on their backs, totally unrecognizable as crustaceans, kicking food into their mouths. Filter-feeding scallops lie on the surface; when they need to flee, they swim with a fast but jerky motion by opening and shutting their shell, like an animated set of false teeth.

Other species that live in the rocky crevices spend their time scavenging. Multicoloured ribbon worms, up to 5 metres long, thrive on the ooze amid the debris. There are flat worms and sea slugs such as the sea lemon, looking more

Five-fold symmetry
A relative of starfish and sea-urchins, the brittle-star (*Ophiothrix fragilis*) feeds by using the minute tube feet on each of its arms to pass food along to the central mouth.

like a mobile omelette than an animal; worms such as the sea mouse and other scale worms; and numerous bristle worms. A vast range of lobsters, squat lobsters and crabs of all shapes and sizes, first seen as larvae in the plankton, settle down here. They include hermit crabs, some draped with an anemone as camouflage, porcelain crabs, spider crabs, swimming crabs, mitten crabs, and edible crabs. Wandering among them are large numbers of starfish and brittle-stars.

In shallow water where the currents are moderate, thick forests of kelp colonize the rock. From above, the kelp resembles a waving mass of leather straps and belts; from within, it is more like a miniature primeval forest, where gnarled holdfasts clutch the rock. A tangle of other weeds and small colonies of hydras, minute relatives of the jellyfish, grows on the forest floor; within the holdfasts lives a community of worms, crustaceans and even small fish, which are found nowhere else in the sea. Moving up and down the weed itself, and grazing upon it, are sea-urchins.

One of the typical fish of rocky shores is the wrasse, colourful in both appearance and nature. Individuals of some species change from female to male as they age. Wrasse form territories, and presumably derive the benefits of an intimate knowledge of their home range, such as the best places to feed and to hide. Other fish found here include blennies, gobies and scorpionfish, all of which eat a wide range of food. But grey mullet, another species abundant on rocky coastlines, specialize in extracting organic material from the ooze, gulping in mouthfuls of mud, roughly sieving it before swallowing. Mullet will also graze on the filamentous algae that cling to rocks and pier pilings.

Dining upside down
The transparent comb protruding from the top of the barnacle (*Semibalanus balanoides*) is actually its feet, with which it takes food particles from the water.

The Kraken
The octopus (*Octopus vulgaris*) glowers from the rocks and stones of the seabed (below). It moves along the bottom using its tentacles, and pounces on its prey.

BIRDS

Birds of many types depend on the North Sea. The cliffs of the north provide a safe haven for hundreds of thousands of seabirds – guillemots, razorbills, puffins, gannets, kittiwakes, terns and fulmars – all drawn to the region by the abundance of fish that thrive at the boundary between Atlantic and North Sea waters.

Some birds, such as cormorants, shags and gulls, along with ducks such as merganser, goosander, eider and shelduck, live in coastal waters all around the sea, although they may move from one area to another during the course of the year.

The numbers of wading birds are swollen for a few weeks in autumn by millions of birds journeying from their breeding grounds in the far north to the warmth of central and southern Africa, where they will spend the winter. They make the return journey in spring.

Another large group of birds, the ducks and geese, also migrate from northern breeding sites, but these stay to spend the winter in the salt marshes and freshwater hinterlands of the North Sea estuaries.

The birds show the North Sea in its true context – as an indispensable part of a global ecosystem, in which each individual sector of the sea plays a vital role.

Foraging for the young
A puffin (*Fractercula arctica*, above) gathers sandeels for its young – its capacious beak can hold as many as ten small fish at once. Puffins breed in large colonies, and lay their eggs in nests underground, often using abandoned rabbit warrens or holes dug by other creatures.

Sea swallows
Common terns (*Sterna hirundo*, right), lured by a shoal of sandeels, descend upon a sandy shore.

Changing the guard
A fulmar (*Fulmarus glacialis*, bottom right) sits tight on its nest while its mate flies above, looking for food. During incubation, which lasts for eight weeks, male and female take turns on the nest, often for several days at a time.

Bird as fish
A shag (*Phalacrocorax aristotelis*, below) transforms from flyer to swimmer in a fraction of a second. Shags breed in small colonies (left), and are common on rocky coasts and islands in the northern North Sea and western Channel.

Avian sieves
A pair of spoonbills (*Platalea leucorodia*, left) preen one another as part of an elaborate courtship. The flat, wide bill is efficient, too, at capturing small fish, molluscs and crustaceans from the water. Spoonbills feed on the mud flats of the Wadden Sea and breed in groups in reed beds.

All of these fish are joined in their search for food by members of the cod family – bib, pollack and whiting – until they themselves are put to flight by a seal. A number of inshore birds raid the larder, too, including ducks such as eider, merganser and goosander, along with gulls, shags, cormorants, divers and terns.

THE TIDAL ZONE

Just beyond the kelp is the intertidal zone, most familiar to us when the tide is out. Then it looks relatively quiet, but the incoming tide brings it rapidly back to life. Now, above the kelp, amid the humped rocks and slopes, a huge array of red and brown seaweeds is suspended in the water, some tough and leathery, some with chalky skeletons, some gristly to the touch, others delicate and leafy.

Nestled among this profusion of weed are animals in frantic activity. Beadlet anemones extend their stinging tentacles and wait for small crustaceans and fish fry to swim into their arms. Shore crabs roam across the surface. On the fronds of kelp and other weeds are the small spiral cases of worms with feeding fans extended, miniature colonies of hydras and other minute animals, with tentacles outstretched to capture the smallest plankton. Limpets glide slowly across the rocks, scouring the thin film of algae with rasping tongues, emulated by small chitons – primitive molluscs that look like a cross between snails and woodlice. Periwinkles crawl over both rocks and weed, feeding on the surface film.

Every stage between low and high tide is marked by a different set of species, each able to withstand exposure to the air for a different length of time. Low down the shore, serrated wrack flows in mops about the rocks. Higher up are bladder and knotted wracks, the former with gas bladders that hold the weed up vertically to the light.

Also in this middle area is a chalky band of billions of barnacles, with shells agape and legs rhythmically kicking, feeding on the vast supply of plankton swilling around in the eddying currents and waves. Barnacles fight a perpetual battle for space with each other and with mussels. As they grow, they lever each other off the rocks to certain death. Spaces left by competition, or by storms, then become available for the next generation emerging from the plankton. Predatory dog whelks drill their way into the shells of barnacles and mussels alike.

Above the middle zone lies yet another species, the spiral wrack; while at the highest level of all, seldom covered by the tides, and out of the water for days at

Life in a rocky pool
Seaweeds share a rocky pool (above) with a ruby red beadlet anemone (*Actinia equina*), topshells (*Gibbula cineraria*) and a rock goby (*Gobius paganellus*).

Flocking to the shore
With the retreat of the tide, a massive flock of waders (above), mainly dunlin (*Calidris alpina*) and knot (*Calidris canutus*), descends on the flats of an estuary looking for food.

Prostrate coils
Thongweed (*Himanthalia elongata*, left) grows best in exposed places, on the lower shore. Its cordlike and elastic structure helps it to withstand the battering it receives from the waves.

a time, lives the small channelled wrack. But even this produces small free-swimming young, which are carried away and live among the plankton before being set down to mature at the top of the tidal zone. Even at the edge of the sea, most species are largely dependent on the phytoplankton food web to feed and disperse their larvae.

The tidal zone of sandy or muddy shores reveals some differences from deeper soft sediments. Single-celled algae, including diatoms, and other minute plants live in the mud, along with such highly visible species as the lugworm. But even if the species differ, many creatures belong to the same general types of families. There are tellins, cat worms and rag worms, a multitude of small crustaceans, and vast numbers of minute and microscopic forms within the sand itself. The one fundamental difference is that there is a new set of predators.

When the tide goes down, a vast army of birds appears, wheeling overhead before descending on the tidal flats, equipped with a lethal armoury of bills variously adapted to chisel, stab, prise, filter, upturn and dabble. There are orange-billed oystercatchers, curlews and whimbrel with their downturned bills, and elegant, long-shanked godwits. Bobbing among these larger birds are flocks of common sandpipers, redshanks, knot, sanderling and dunlin. For them the soft sediments are a table replenished with every tide.

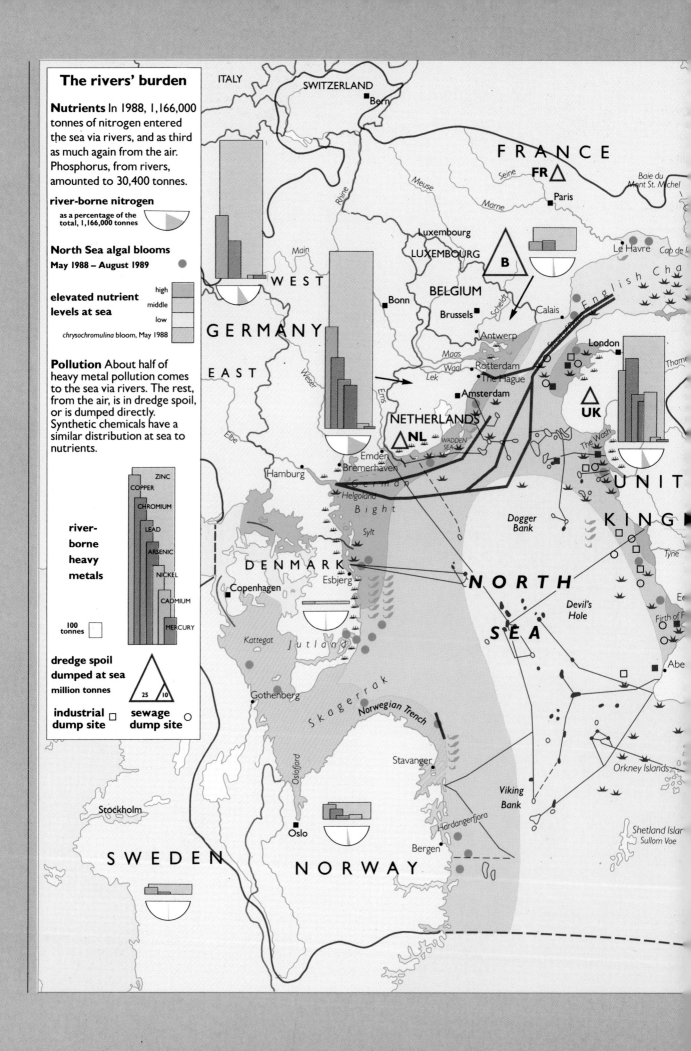

PART 2

Troubled sea

Off shore industrial pollution is increasing. In 1988 30,000 tonnes of oil were lost.

oilfield   gasfield   pipe

Shipping separation schemes have not prevented accidents or pollution. In addition, up to 60,000 tonnes of oil are dumped annually.

separation schemes

The military dumps weapons and uses the sea for exercises.

dump ■   range

Habitats such as estuary mud flats and coastal saltmarshes remain under threat.

20 km² saltmarsh

Fishing Size of national catch in the mid 1980s.

100,000 tonnes landed

Islands (mark)

People first arrived on the shores of the North Sea towards the end of the last ice age, 13,000 years ago. For 12,500 years we had little effect. Then we began to find ways of taking land from the sea, helping to spark off an agricultural revolution, but destroying much of the coastal wetlands in the process.

In the late 19th century, we began to use larger fishing vessels and ever more sophisticated fishing techniques, until we took more fish than the sea could bear. On land the industrial revolution produced pollution, which made its way via the rivers into the North Sea. Over the years, an increasingly complex cocktail of highly toxic chemicals has been dumped into the sea, while our energy-hungry way of life, along with intensive farming, has overloaded the sea with nutrients. In the last 25 years a new source of pollution has developed following the discovery of rich reserves of oil and gas under the seabed. A vast trade in raw materials is now carried across the North Sea by ship, and accidents, as well as pollution from the ships themselves, have added their burden to the sea.

Assaulted from all sides
When the North Sea is seen from a new perspective, its true form is revealed: it is a bay on the northwest coast of Europe, opening into the Atlantic in the north, with a narrow channel into the ocean to the south. Pollutants and nutrients are not rapidly flushed out into the wide ocean. Instead they linger in a sea that has many other demands placed upon it by the people and industries clustered around its shores.

LOST HABITATS
THE TAKING OF LAND AND SEA

THE WETLANDS that line the shores of the North Sea – mud flats, brackish lagoons, salt marsh and waterlogged hinterlands – rank equally with tropical rainforests and coral reefs as the world's most productive biological systems. Each year at least 2 kilograms of carbon are captured beneath every square metre and turned into organic material. This profusion is encouraged by the nutrients that are brought down to the coast by the rivers, and by the organic debris brought from the sea by the tides and currents.

Life flourishes here: single-celled algae thrive in the shallow, sunlit water and on the surface of the mud, while sea grasses grow in dense beds just below the tide line. Beneath the quiet surface of the mud flats and saltmarsh creeks is a seething mass of many thousands of worms, small crustaceans and molluscs, which attract the millions of wading birds that feed on the shores of the North Sea, and provide a rich source of food for fish when the tide is in. The salt marsh that surrounds the creeks is rich in plant life, too, including glasswort, sea lavender and grasses; these support great numbers of insects, molluscs and other small animals, all of which draw in ducks, geese and many smaller birds.

Yet, over the centuries, these wetlands have been under increasing pressure, as more and more have been drained to create land for agriculture and industry, and barricaded to provide defences against the sea.

Wetland wilderness
Sheltering behind the coastal sand dunes are the tidal creeks of Blakeney Marsh in East Anglia, part of the largest salt marsh on the North Sea's west coast. The swards of sea lavender produce copious quantities of nectar and, in summer, hum with the sound of bees. Mixed in with the lavender is a unique combination of other flowering plants, grasses and rushes. Breeding birds such as the redshank and curlew feed on the insects and other animals living among the vegetation, and on the many small crustaceans in the mud of the warm shallow creeks.

From wetland to farmland

Coastal communities have always waged a battle with the sea. On the North Frisian tidal flats, in the German Wadden Sea, can be seen the remains of ditches, dykes and paths, and of the raised platforms of houses that were abandoned as long ago as the Middle Ages. On parts of the east coast of the United Kingdom the cliffs are still eroding, at more than 2 metres a year, and whole villages have been lost as a result. When the sea had the upper hand, rivers were tidal for 40 kilometres or more, and they meandered across flood plains, stretching back to huge areas of mud, marsh and fen, where salt water blended to fresh.

But the balance of power between people and the sea has changed over the centuries. In fact, reclamation of land at the edges of the Wash in eastern England started in Saxon times, and the size of the Wash has virtually halved since then. Even as the Frisian farmlands were slipping beneath the sea, drainage engineers in the Netherlands were working to improve the methods of impounding their coastal waters and hinterlands. Despite the massive drainage programmes of the 17th and 18th centuries, many stretches of marsh survived; but by the late 19th century even these existed only as memories, of a world where 'high overhead hung, motionless, hawk beyond hawk, buzzard beyond buzzard, kite beyond kite, as far as the eye could see.'

The newly created water meadows did at least provide a habitat for some waders, while the drainage canals contained a fairly diverse range of rushes, waterweed, dragonflies and newts. But from the 1950s increasingly intensive farming methods demanded that the fields be drained and the vegetation controlled with herbicides, creating barren flats and dykes empty of life. In extreme cases, the cattle and sheep that once grazed the land were displaced by cereal crops. Wildlife became restricted to reserves. Where once the redshank strode and the ruff displayed, only the wind remained.

Despite the 'set aside' schemes that are taking farmland out of production to reduce surpluses of grain and other crops, land claims for agriculture continue. Between 1979 and 1981, the Højer sluice project on the Danish/German border of the Wadden Sea went ahead despite the fact that it would destroy a valuable natural habitat, by the building of an embankment between two headlands in order to drain an area of mud flats and salt marsh. But protests by environmental groups led to part of the land being set aside to create a brackish marsh and lagoons, and this has enhanced the numbers of certain waders, ducks and geese drawn to the area. Nevertheless an even larger area of natural habitat has been lost.

No through road
The loose deposits of boulders, gravel, sand and clay left by the last ice age in eastern England are rapidly being eroded by the sea. In some areas the coast has retreated 4 kilometres in 2,000 years.

End of the line
Natural habitats in coastal areas have been greatly modified by humans. The salt marsh at this site (below left) in Altenbruch, Germany, was turned to farmland long ago. Now, this too is being destroyed, to be replaced by an industrial complex.

Havergate Island
The loss of natural habitat is so great that some areas such as this bird reserve at Havergate Island in the UK (below) are managed to provide as diverse a wetland site as possible. Sluices regulate the water levels, and the lagoons and pools are excavated to different depths.

Damming the sea
The Zuider Sea was dammed in 1932 with a 28-kilometre barrage (left). This was built with a core of sand dredged from the Wadden Sea, and waterproofed with a covering of clay taken from the Zuider Sea.

Self destruction
Channels in the mud flats mark a reclaimed area of the Wadden Sea. Mud deposited in the channels by the tide was thrown on to the developing fields. The process extended ever further seawards – thus the sea became the agent of its own withdrawal.

Even after an area has become a national park, natural habitats are not safe, as demonstrated by the enclosure, in 1987, of the Nordstrander Bucht in the German Wadden Sea. Elsewhere, huge drainage schemes are still being put forward, and small habitat creation projects are proposed as part of these, but in reality they are merely palliatives offered by the developers.

Draining the sea

One drainage scheme has surpassed all others, both in the scale of its ambition, and in its impact on the North Sea. In 1932, 2,700 square kilometres of the shallow, inland part of the Wadden Sea – the region known then as the Zuider Sea – were cut off by a 28-kilometre-long dam. The salinity of the enclosed area – renamed the IJsselmeer – dropped rapidly: marine life died, to be replaced by freshwater species. The Zuider Sea herring, unique to the region, were among the casualties, as were their predators, the bottlenose dolphins, which disappeared from the Wadden Sea altogether. The failure of the Wadden Sea eel grass to recover from a disease that swept the North Sea in the 1930s is thought to have been caused by the change in currents and sedimentation that followed the damming. As the eel grass disappeared, two species of fish associated with it became extinct in the Netherlands; the populations of other species that graze on eel grass, such as the brent goose, also declined.

Within the IJsselmeer, huge polders have been created by pumping out the water to create an area of land up to 5 metres below sea level. Some 70,000 tonnes of pork are now raised on this land every year, replacing an average catch of 12,500 tonnes of herring. Although this may be seen as a triumph and justification for the reclamation, in reality it represents a twofold loss for the North Sea: not only has a vast expanse of sea vanished for ever, but the pigs are reared on great quantities of fish meal – a product of the industrial fisheries that are currently exploiting the waters of the North Sea.

One more polder is planned for the IJsselmeer: a 41,000-hectare area known as the Markerwaard. But with the growing awareness of the damage it is likely to cause to the environment – and of the fact that there is no need to create more land for agriculture – the reclamation now seems unlikely. In any case, the same amount of farmland could be gained in only three years by stopping the development of green fields and building instead on derelict urban land.

Burying the estuaries

Another threat to the wetlands, particularly those in estuaries, is the reclamation of land for industrial development or waste disposal. And the pace of destruction is accelerating, as a result of new techniques that allow embankments to be constructed in deeper waters and more exposed areas than before.

The development of the ports of Rotterdam and Antwerp involved huge construction works, and had correspondingly devastating effects. But the cumulative effect of numerous smaller developments is just as damaging. Often they are allowed because the planners believe that an estuary is nothing more than a mass of mud, or that the wildlife can always move elsewhere.

But there is nowhere else for the wildlife to go. Other sites are already occupied to capacity. When birds are crowded together, the flurry of activity alerts the shellfish, worms and crustaceans on which they feed, and the prey retreats deep into the mud. In winter, wading birds have to eat the equivalent of a third of their body weight each day; anything that makes this more difficult has fatal results.

One case, that of Fagbury Flats in the Orwell estuary in England, is typical of many throughout the North Sea states. Fagbury was the roosting place of 5,000 waders, and the feeding ground of 3,000, including threatened species such as redshank and dunlin, whose numbers have declined by between 22 and 45 percent along the west coast of the North Sea since the 1970s. Yet the flats disappeared under the concrete of a new container port in 1989 – even though existing ports had spare capacity, or could be expanded with fewer harmful effects. Conservation groups lost their battle to preserve the area because the development company wielded great political power and received support at all levels of government – and because, in the UK, there is no system for weighing the advantages of coastal development against the cost to the environment. Even in countries that do have such systems – Norway, Sweden and the Wadden Sea states – the difference between theory and practice is often large.

The sea subjected
With today's methods and machinery, developers can claim huge areas of the sea at once, as they have at Altenbruch. Industry prefers to develop rural sites because the costs are low, and because there are often financial incentives to provide employment in such areas. Sites on the coast once offered an easy means of waste disposal – directly into the sea – but this should now be coming to an end.

Birds in peril
A flock of knot (*Calidris canutus*) crowds on to an island at high tide. When industry claims areas of mud flats in the North Sea, it cannot assume that the large population of wading birds that depend on them will simply find somewhere else to go. When a stretch of land is developed, the birds displaced are likely to die.

The best defence?
Concrete blocks, such as those used in this coastal defence at Helgoland, Germany (above), may protect the coastline, but they provide little scope for marine life.

The art of restoration
Eroding sand dunes can be repaired by erecting fencing (right) and allowing the sand to settle.

Recreational wreckage
A crowded beach is a dramatic example of the complete destruction of a natural habitat. Where once birds would feed at the water's edge, now only bathers line the shore. But habitat destruction can be more subtle than this: even a single person walking along the beach at high tide may disturb thousands of birds, and one or two disturbances a day may ruin the value of a site for wildlife.

In some estuaries – the Tees, in England, is one example – 90 percent of the mud flats have been given up to industry. Yet there is no shortage of derelict industrial land that could be brought back into use; indeed, this must happen if those estuaries that remain are to be protected.

Barricading the coasts

Since the retreat of the ice 13,000 years ago, the North Sea coast has been in a constant state of flux, and considerable effort has been devoted to protection schemes – both to prevent flooding and to stop the coast eroding.

In order to hold a particular boundary against the sea, often one that encompassed huge areas taken from the sea itself, and now totally alien from it, coastal engineers resorted to the use of 'hard' artificial structures such as dykes above the tide line, or concrete blocks or sheet-steel pilings in the water. In the worst cases, material was taken from the shore and dumped behind the barriers, creating a sheer, lifeless cliff between land and sea, where once stretched hundreds of metres of marsh and mud.

Now, however, the engineers seem to be increasingly aware of the need to work with nature, to use the 'soft' natural defences of dunes, salt marsh and shingle beach. But still there are difficulties. Sand dunes, for instance, can be destroyed extremely easily if large numbers of people walk on them, breaking the protective skin of vegetation. Once this has happened, the sand blows away in the wind. This kind of damage can be repaired if the dunes still have a supply of sand, washed up by the sea. But some dune systems no longer receive fresh sand – either because the offshore deposits are exhausted, or because the currents that once brought the sand to the beach have changed as the shape of the coast has altered – and these sites are particularly vulnerable to erosion.

GLOBAL WARMING AND THE NORTH SEA

So far this century, sea level has risen by 10–15 centimetres. The pace is expected to accelerate, with a rise of another 20 centimetres by 2030, and a further 60 centimetres by 2100. The level may even rise by as much as 1 metre by 2040. Low-lying areas such as the islands of the Wadden Sea, the whole of the Belgian coast, and parts of the east coast of England may be inundated.

The rise in sea level is just one of the consequences of the warming of the global climate, which itself has been caused by so-called 'greenhouse' gases – mainly carbon dioxide and methane – trapping heat close to the Earth's surface. The level is rising because water expands as it gets warmer, and because the polar ice is melting.

As a result, conditions in the North Sea are expected to alter. As well as the water becoming warmer, the currents and the patterns of the tides will shift, resulting in local changes in salinity and turbulence. And, without doubt, wildlife will be affected as the physical nature of their habitats changes.

The consequences for tidal flats, such as the Wadden Sea and the Wash, depend on a number of factors. Researchers believe that the rate of sedimentation could just keep up with a rise in sea level of around 1 metre in 100 years; but if it does not, the flats will sink below the water. The many unique plants of the salt marsh appear to be even more at risk – they can grow upwards at the rate of only 50 centimetres per century. Moreover, the retreat inland of the salt marshes, ahead of the rising water, will be blocked by fixed sea defences. Similar problems will be created in estuaries.

More than 50 species of ducks, geese, waders, gulls and terns – both breeding and migrating birds – depend on the Wadden Sea alone. They are already under pressure because their feeding areas are being developed for industry and agriculture, and any further loss would harm them beyond measure. Much the same can be said for other, similar areas around the North Sea – the Humber, Wash and Thames, the Solent and the Baie du Mont St Michel.

Fish in the North Sea would also be affected by the change in climate. Many species use the warm, shallow waters as nursery grounds; but if the currents alter, the fish eggs and larvae might be washed into deeper and colder areas where they would not be able to survive. The North Sea also acts as a boundary between warm and cold water fish. Species at the northern limit of their range include sardine, mackerel and anchovy, while cod, haddock and herring are at their southern limit. Temporary shifts in distribution already occur because of fluctuations in the climate, and a permanent warming would result in major changes, whose effects would be felt throughout the food web.

There are some things that can be done to reduce the impact of a rise in sea level, particularly by giving coastal habitats room to move inland. Withdrawing the dykes, for example, may provide some scope for retreat. Other changes, however, such as the alteration in distribution of fish, could not be prevented.

The conclusion is that for the North Sea, as elsewhere, the problems caused by the warming of the climate are potentially so great that emissions of the gases responsible must be drastically reduced. The complexity of this task is all the more reason for action to be taken now, before marked changes occur.

Island survey
The island of Vogelsang in the German Wadden Sea encapsulates some of the problems that can be expected from the rise in sea level that is likely to take place as a result of the Earth's climate becoming warmer. Natural habitats such as the salt marsh to the right will retreat inland, unless halted by sea defences. The sandy beaches on the left will also become smaller, to the detriment of both the wildlife and the people who use them for recreation. Another problem to be faced by the island's inhabitants will be the possible contamination of drinking water as salt rises in the water table.

The killer sea
In the winter of 1953, gales created a huge storm tide, which breached the sea defences and flooded 2,000 square kilometres of the southwestern Netherlands, claiming the lives of 1,843 people.

Completion of the Delta Plan
The closure of the Eastern Scheldt in 1978 was designed to prevent a recurrence of the 1953 floods. If it were planned today, more might be done to reduce its effect on the environment.

Salt marshes offer remarkable protection against the sea, reducing the force of storm waves by 90 percent, and their height by 70 percent. Yet much of the salt marsh around the North Sea has been converted into farmland, and that which does remain is eroding. Once in decline, salt marsh is very slow to regenerate.

In the case of shingle beaches, erosion can be halted by adding more shingle, or by building groynes to stop it drifting too far along the coast. Even this seemingly simple operation needs to be carefully managed, since trapping shingle at one point is likely to starve another area further along the shore – and this, in turn, may start to erode.

The low-lying Netherlands have faced the most difficult problems, and here artificial defences have been seen as the only answer. In the disastrous flood of 1953, in which more than 1,800 people lost their lives and huge areas of land were inundated by the sea, the weak point was the complex of rivers and islands in the southern Netherlands that made up the North Sea's only large delta. Around these shores ran some 700 kilometres of dykes, all of which would need to be strengthened or rebuilt if a repetition of the tragedy were to be avoided. Instead, engineers implemented a 25-year-long 'Delta Plan', erecting a series of dams across five of the seven inlets in the delta. This cut the length of the coastal defences to 30 kilometres, but it also destroyed the natural habitat in the process: the salinity of the water changed, currents and tidal range altered, and levels of oxygen decreased – all of which had a drastic effect on the natural inhabitants.

In the mid 1970s the final stage of the plan, the damming of the Eastern Scheldt, brought a storm of protest from fishermen and environmental groups concerned at the threat to the river's particularly diverse marine life. There were important mussel, cockle, shrimp, eel and crab fisheries; more than 200,000 waders and ducks flocked there during the winter months. Tidal barriers were built instead of the dam, with assurances that the local ecosystem would then be

The end of the Rhine
The northern mouth of the Rhine, which enters the North Sea west of Rotterdam (right, foreground), provides an extreme example of the modifications humans choose to make to nature. The river once freely shifted its course through the sand dunes, but is now imprisoned between walls that extend out into the sea. The estuarine life that manages to cling on there is a travesty of what went before. The extension of the canal out to sea also blocked the flow of coastal sediments, which now pile up against the canal wall.

The once and future scene?
Sand dunes and salt marsh were once a common sight along the coast of both the English Channel and the North Sea. It is within our power to help nature to return once more.

perfectly safe – but the barriers still block most of the estuary mouth in normal conditions, and the effects have been far greater than anyone had predicted.

Since the barrage was completed in late 1986, nearly 60 percent of the salt marshes in the area have been lost, along with 45 percent of the mud flats, with a commensurate loss of the birds that used to feed there. This highlights how little anyone understood of the ecology of the region, and emphasizes the need for far greater caution in planning any scheme of this kind.

The Delta Plan came from an age when the main problems were seen as technical ones – the environment did not merit consideration. Were the scheme to start now, a different approach might be taken, such as installing a series of barriers that swing up out of the seabed only when needed, as the Thames barrier does. There is, in any case, a need to reassess the necessity of coastal protection, especially in those areas where there is no danger of flooding or where its extent can easily be limited. It would be better to protect only vital sites from erosion – a barrier that protects a town, for example – and let others revert to a more natural condition. At the mouth of the Elbe, the withdrawal of sea defences inland is now being seriously considered, to allow the area between them and the sea to revert to its natural state. If this were to happen on a broad scale, then perhaps the sight of buzzard beyond buzzard, kite beyond kite, could once more become common.

FISHING
THE RELENTLESS HARVEST

THE NORTH SEA is one of the world's major fishing grounds. Although it represents less than one fifth of one percent of the world's seas, more than five percent of fish caught every year come from its waters – even after many years of intensive fishing. During the 1970s more than three million tonnes of fish were taken each year, and although this number has since declined to less than two and a half million tonnes, every year it still represents a staggering 25 percent of the fish estimated to be living in the North Sea.

There are two main types of fishing for human consumption: fishing that takes place in surface waters – known as 'pelagic' – and that which occurs at greater depths – called 'demersal'.

The main pelagic species are herring and mackerel. Adult herring congregate wherever the plankton on which they feed are plentiful. In the breeding season different groups move to spawn in distinct coastal regions, mainly in the western North Sea, and currents then sweep the young larvae and immature fish into the German Bight and adjacent areas. Some shoals of mackerel also live and breed in the North Sea, but others simply pass through its waters and breed elsewhere. Since the 1950s the main gear used for pelagic fishing has been the purse seine.

The prime target of the demersal fisheries is the cod family – particularly cod, haddock, saithe (coley) and whiting. These species are separated geographically, to an extent: haddock are more abundant in the north, cod in the south; whiting prefer coastal waters. Different species also feed from different types of seabed – sandy, muddy or rocky – while some, such as whiting, will be drawn even to old wrecks. The cod family are caught mainly using otter trawls and seines.

Working in pairs
Two trawlers working together can pull a larger net than a single vessel, and catch many more fish.

Easy prey
The shoaling habit of herring may afford them protection from other marine creatures, but shoals are easily found – and caught – by fishermen.

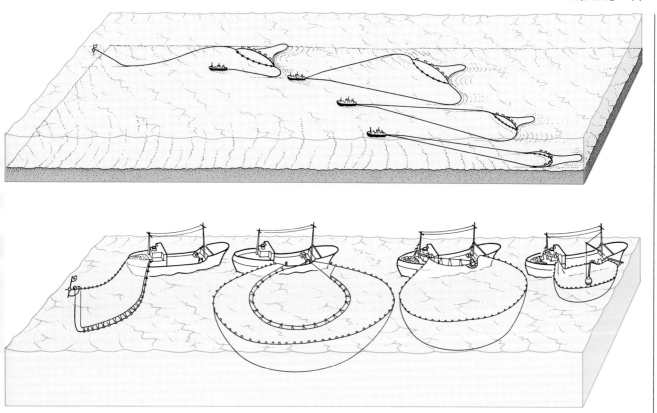

Bottom-living fish, particularly plaice and sole, constitute the other major group exploited by demersal fishing. Like the cod family, the different species are dispersed around the North Sea; and some, such as turbot, provide small but very valuable catches. Over half of the plaice, the most important species, are taken by the Dutch fishing fleet. This kind of fish is caught with beam trawls, which cause an enormous amount of damage in the sea: their chains drag along the seabed, breaking up the surface and greatly disturbing the creatures that live on it. In areas where fishing is most intensive, such as the Wadden Sea, barely a centimetre of the seabed remains intact.

There is a second kind of fishing in the North Sea – the so-called industrial fisheries – which grew up at lightning speed from the late 1950s. Industrial fishing centres on fish of all species that are considered too small for direct human consumption. They are rendered into fish meal and oil for use in foodstuffs in intensive livestock units, fish farms and mink farms, as well as in pet food.

Species taken by the industrial fisheries include Norway pout and blue whiting, sandeels and sprats; however, young herring, whiting and haddock may also be caught. Most industrial fishing now takes place in the north, although fishing for sandeels is restricted to shallow waters, where the fish burrow into the seabed.

The main fishing nation in the North Sea is Denmark, which in the mid 1980s was responsible for about half the catch, most of it industrial fish. Norway took some 20 percent of the total, just under half of which consisted of industrial fish; and the United Kingdom took 15 percent, a tenth of which was industrial. Other North Sea states that fish there are the Netherlands (7 percent), France (4 percent), West Germany (3 percent) and Belgium (1 percent). Other countries do fish in the North Sea, but their total catch is always less than 1 percent.

FISHERIES MANAGEMENT

Both the capacity and the efficiency of the North Sea fishing fleets have been growing steadily for many years, along with the number of species hunted by them. Indeed, there has been concern about the impact of fishing in the North Sea

Seine fishing
In deep-water seine netting (top), after first encircling the fish with a net, the ends of the rope are hauled together, herding the fish into the bag end. In surface purse seining (above), the bottom of the net is gathered before the net is drawn in and the fish are pumped out.

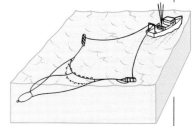

Otter trawling
The ship tows a bag net, kept open by otter boards.

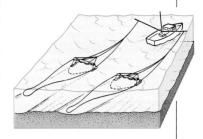

Beam trawling
The net is hung on a rigid frame and dragged along the seabed.

Samples for science
Fine-mesh nets (above) enable scientists to take samples of the plankton and to monitor the numbers of fish eggs and larvae present at any one time. Armed with this information, they attempt to predict the populations of fish in the coming year – with varying success.

Gross harvest
Norwegian fishermen (below) haul in a purse seine overflowing with herring and remove the fish using a large scoop net. The herring fishery in the North Sea was so huge in the 1960s and 1970s that the herring population collapsed.

since at least the end of the 19th century when, with the arrival of steam power, there was a dramatic increase in the harvest from the sea. In some major ports, fishing increased 400 percent in the space of 30 years.

As a result, governments began to be concerned that certain species would be fished so heavily that their populations would fall drastically, to such a level that there could be no profit in fishing for them. If governments were to prevent this, they would have to be able to predict what would happen to different species over a period of years, so that they could set target figures for each catch.

As a first step, biologists decided that by monitoring the catch in any one year, they would be able to estimate both the total population of each species and the number of eggs it would produce that year. In this they were fairly successful. They also hoped to be able to predict how many of these eggs would survive to become adults. Once they understood the complete life cycle they would be able to achieve their goal, of making a long-range forecast of how many fish there would be – and therefore how many fish they could safely take.

But the number of fish larvae joining the adult population turned out to vary enormously from year to year. This could have been for one or more of a number of reasons, including the amount of food available, the prevalence of disease, the threat from predators, or the effect of storms and currents taking the young far from their home waters. But hard information was difficult to gather, and biologists could only make a guess at what effect any of these factors might have and, having done so, go on to draw up mathematical models of what would happen to future fish populations so that they could manage the fisheries. When these models were first produced in the late 1950s, they appeared to hold out some promise – but 30 years of research have not greatly improved their accuracy.

Because the models are inaccurate, short-term forecasts of the population have to be made. A survey of immature fish is carried out a few months before they

enter the adult stock, when it is fairly certain that most will survive. This gives a more accurate idea than the models – but only for the year ahead. Despite these difficulties, governments have continued to use both models and forecasts to regulate the North Sea fisheries, ignoring warnings of their limitations.

The main aims of fisheries management have always been economic and social. The pursuit of profit has taken priority over the protection of the fish; and the reluctance of the fishing industry to accept restrictions, coupled with the lack of political will to take effective measures in time, has had inevitable results.

FISH STOCKS IN THE NORTH SEA

For want of proper regulation, the pressure of modern fishing methods became too much for the fish population to bear long ago. First the surface herring fishery collapsed in the 1960s. The bottom-feeding stock – the fish that live in deep water and on the seabed – largely collapsed in the 1970s and 1980s; and now the industrial fishing stock shows every sign of going the same way in the 1990s.

Fishing in surface waters

Even as the herring population plummeted in the 1960s, new methods such as purse seining and echo location meant that shoals of herring could still be found. The stock suffered further when the Norwegian purse-seining fleet, having cleared out the herring to the north, moved into the North Sea. The spawning population of herring, thought to be around 2 million tonnes in the early 1960s, fell rapidly as over 60 percent of the stock was removed each year. When herring fishing was prohibited in 1977, only 60,000 tonnes of spawning fish remained. In 1983 the population had partially recovered and fishing began again. But the scientists' recommended catch of 98,000 tonnes was increased by politicians to 145,000 tonnes – and the actual catch was 308,000 tonnes. This pattern became typical, and in 1988, once more, the stock stopped increasing.

The mackerel fared even worse. Mackerel are long-lived fish that breed relatively slowly. When the herring fishery collapsed in 1967, the purse-seine fleet transferred most of its attention to mackerel, and this stock, too, rapidly diminished. Mackerel enjoyed one last population boom in 1969, when an extraordinarily large number of fish survived to become adults, but since then their numbers have been dwindling. When the last remnants of the 1969 population disappear, biologists believe the North Sea mackerel will probably become extinct, leaving only visitors from the Atlantic in the northern waters.

Deep-water fisheries

The cod family and various species of flat fish make up most of the deep-water fisheries. In the 1960s the spawning stock of the cod family – haddock, cod, saithe and whiting – unexpectedly increased between two- and fourfold, depending on the species and, during the 1960s and 1970s, there were record landings of all four species. The fishery has remained intensive ever since, even though the fish have long stopped breeding at the same rate.

The proportion of the haddock population caught today is greater than at any other time on record. Fish are pulled out of the water within a year or two of reaching spawning age when, in natural circumstances, they would live and breed for two decades or more. The last passably good year for young reaching spawning age was 1983, but few of these fish are still alive. The spawning stock is now the smallest it has ever been.

The story of cod is almost identical. The spawning population has decreased to its lowest level in 30 years or more, and the number of young fish caught (particularly in the German Bight) has been growing – leaving even fewer fish to breed new generations. In 1987, landings of cod were the lowest for 20 years.

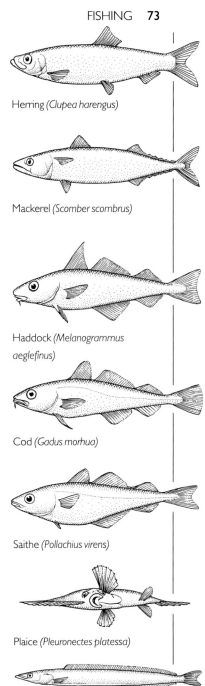

Herring (Clupea harengus)

Mackerel (Scomber scombrus)

Haddock (Melanogrammus aeglefinus)

Cod (Gadus morhua)

Saithe (Pollachius virens)

Plaice (Pleuronectes platessa)

Sandeel (Ammodytes tobianus)

A price on their heads
Species of fish that are important commercially include herring and mackerel from the surface waters, cod fish such as haddock, cod and saithe from the middle reaches, and bottom-feeders, including plaice. Sandeels have become the mainstay of the industrial fisheries.

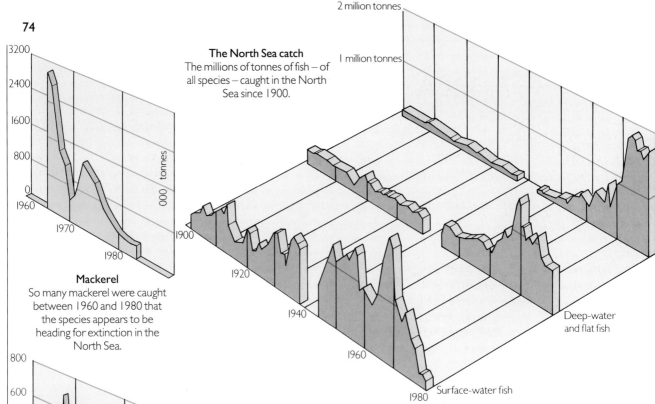

The North Sea catch
The millions of tonnes of fish – of all species – caught in the North Sea since 1900.

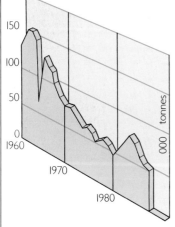

Mackerel
So many mackerel were caught between 1960 and 1980 that the species appears to be heading for extinction in the North Sea.

Haddock
Stocks are already low, but poor breeding in 1987 and 1988 means that their situation is expected to become even worse in the 1990s.

Sole
Levels of stocks were low in the late 1980s. If the fish have a good breeding year – as they did in 1980 – they may recover.

Saithe has not yet been fished as intensively as cod or haddock, but an increasing number of fish are caught, and again, in the late 1980s, the spawning stock was among the lowest on record. Whiting is the only species that has not plunged into decline. Although comparatively few fish survived to become adults in the early 1980s, their numbers appear to have been restored between 1985 and 1987. Even so, no one knows whether this is because whiting have been under less pressure, or simply because they have had a lucky run of bumper years.

Flat fish

Plaice and sole are the two main species of flat fish still caught commercially in the North Sea. In fact, plaice appears to be the only major commercial species so far that has survived the general onslaught. The annual catch grew from 100,000 to 150,000 tonnes between the 1960s and 1980s. During the first part of this period the size of the catches was maintained only by investing heavily in technology. Like the herring, the total plaice population rapidly declined. But in the 1980s, plaice unexpectedly increased their numbers, and the threat receded a little.

The Netherlands takes 80 percent of all North Sea sole, having constantly re-equipped its fleet with more sophisticated equipment since the early 1960s. But sole breeding has not improved, and it has been calculated that the population has fallen by two thirds during this period, from 150,000 to 50,000 tonnes.

Industrial fisheries

Industrial fishing grew at a phenomenal rate from the late 1950s, reaching its peak in the 1970s. Some industrial fish are simply catches intended for human consumption that turn out to be surplus to requirement or too small for sale. But the major part of the industry consists of smaller species that, by and large, are now being fished for the first time. These include sandeels, small members of the cod family such as the Norway pout and blue whiting, and the sprat. The fine mesh of the nets used to catch them also traps immature fish of larger species such as herring, haddock and whiting – reducing their populations, and their chances of recovery to previous levels, still further.

In 1974 the total catch of industrial fish was 1,857,000 tonnes. By 1985 it had almost halved, to 1,033,000 tonnes. By this time sandeels accounted for more

THE THREAT FROM THE FISH FARM

The farming of fish has become a boom industry over the past decade and is seen as one way to relieve the pressure on fish in the wild. In the North Sea area, fish – notably salmon – are farmed in Scotland and in Sweden, but the greatest success has been in the sheltered fjords of Norway: in 1989, more than 700 fish farms produced 80,000 tonnes of salmon. Yet, the more the industry grows, the more problems it creates for the environment.

The fish are kept in enclosed pens or cages, moored to the seabed, and fed high-protein food to encourage rapid growth. But for every 10 tonnes of food added to the cages, around a tonne of waste sinks to the seabed, and another half tonne of soluble nutrients dissolves in the water. In some fjords and sea lochs where water circulation is poor, the presence of additional nutrients has resulted in spectacular blooms of algae, some of which have been toxic to marine organisms.

To keep disease at bay, various chemicals are added to the water, including chlorine, sodium hydroxide, iodine, calcium oxide, formaldehyde, copper salts, antibiotics and pesticides, one of which in particular – the pesticide Nuvan – causes considerable problems. Nuvan contains dichlorvos, and is used to kill the parasitic salmon louse *Lepeophtheirus salmonis*; but the dichlorvos washes out of the cages into the sea, killing shellfish, crabs, lobsters and shrimps. Dichlorvos has also been blamed for a huge increase in the number of wild salmon in Scottish waters that have eye cataracts.

And it is the threat that farmed salmon pose to wild fish that gives the greatest cause for concern. The escape of the salmon fluke *Gyrodactylus salaris* from Norwegian farms has devastated wild populations, and has spread rapidly to at least 33 Norwegian rivers. Another fear is that farmed and wild fish will interbreed, altering the genetic make-up of the wild fish and making them less able to survive. Farmed salmon are bred to grow fast, reach sexual maturity late (so that they devote their energy to producing flesh rather than to reproduction), and to be less aggressive than in the wild. (Surprisingly large numbers of salmon escape from farms: in 1987, a survey of 54 Norwegian rivers found that 15–20 percent of the fish were escapees.) Some say there is no evidence for interbreeding; others argue that by the time proof is available it will be too late to act. For once the interests of the farmer and of the environment coincide: for the benefit of both, farmed fish should not be allowed to escape into the wild.

The greatest irony is that salmon farms are dependent on industrial fish for food. Every tonne of salmon produced requires 3.3 tonnes of sandeels or other industrial species to be processed into pellets. In other words, the 1988 Norwegian crop of salmon required one quarter of a million tonnes of fish to be taken from the sea. The idea that farming will relieve pressure on the fish in the North Sea is, therefore, a myth. The only way in which some relief might come is if we acquire a taste for vegetarian species such as grey mullet.

Brief lives
Salmon crowd to the surface as the nets in their pen are drawn in and lifted for harvesting.

Accidental death
A haul of industrial fish may often contain a 'bycatch' of potentially more valuable species: this sandeel catch contains a significant proportion of young haddock. There are regulations that set acceptable levels for a bycatch, but they are rarely enforced.

than half of the industrial fish caught in the North Sea each year. Sandeels have a natural life span of five or six years. Industrial fishing methods have reduced this to just two – making it impossible to predict future population sizes.

The sandeel once shared the burden of the industrial fisheries with two other species, the Norway pout and the sprat. The populations of both have now crashed. Today there are so few sprat that it is no longer possible to carry out an accurate survey of their numbers. Given the collective experience with haddock, cod, saithe and sole, and the difficulties of managing such a short-lived species as sandeels, it is likely that their population will also collapse during the 1990s.

EFFECTS ON THE ECOSYSTEM

There is no doubt that the fishing industry has devastated individual species in the North Sea – and it may well have had a profound effect on the entire ecosystem, but until now there have been few systematic attempts to find out. Only in 1990 did the North Sea states make a clear commitment to examine closely the likely consequences of intensive fishing on life in the North Sea.

Interactions between species

Efforts to predict the different effects that fishing for one species has on others have foundered – because we simply do not have enough information about the interactions between different kinds of fish. In 1981, an 'International Stomach Sampling Project' was carried out in the North Sea to attempt to find out more. Biologists examined the stomachs of 40,000 fish representing five major predatory species – cod, haddock, whiting, saithe and mackerel – and discovered that between them these species eat the same weight of fish as is taken by the fishing industry. So it could well be that the predation of one species on another plays a significant role in controlling population levels – and that reducing or increasing the numbers of one disrupts the natural balance.

The business of predation may be the link between the drastic decline of herring in the late 1960s and 1970s and the massive increase in numbers of the cod family. Because there were so few adult herring left to prey on larval fish among the plankton, it is possible that more cod survived to become adults. Other species that live among the plankton and whose populations have boomed, such as plaice, may also have thrived because of the lack of herring, as may sprats, which as adults might be expected to compete with the herring for food.

All flesh is fish?
Sandeels are sucked out of the hold of a ship, to be taken for processing into fish meal. This high-protein food is fed to fish, animals and birds reared using intensive methods, and it is also used in pet food. Animals destined for human consumption have to be taken off feed based on fish meal for the last part of their lives, otherwise the flesh will taste fishy.

The impact on wildlife

Exactly what impact the fisheries have had on the North Sea ecosystem as a whole is not known for sure. But it seems highly improbable that the birds, seals and cetaceans, and all the other species that share the sea should live independently of one another, and remain totally unaffected by what happens to others.

One might expect that when so many major fish populations crashed, in the 1960s and 1970s, other forms of wildlife, too, would have been severely affected. But, paradoxically, seabirds were thriving at this time – gorging on the huge amounts of waste thrown overboard from fishing boats. Perhaps seabirds and seals benefited because industrial fish themselves became more plentiful as the larger, predatory species were taken from the sea for human consumption.

But the disappearance of the larger fish probably did affect the cetaceans – the dolphins and porpoises, which depend almost entirely on them for food. It is generally agreed that there has been a long term decline in the abundance of most cetaceans in the North Sea and that the most likely reasons for this are the combined effects of fishing and pollution.

It may be only when industrial fishing, particularly of sandeels, collapses on the scale of the herring fishery that significant effects on seabirds will become unmistakable. The experience in the seabird colonies in Shetland could well be a taste of what is to come.

In 1984, 1985 and 1986, arctic terns reared very few young; in 1987 and 1988 the colonies reared no young at all. The same is true of kittiwakes: in 1986, only one chick in every two nests survived; by 1988 this figure was down to one fledgling every ten nests. The 25,000 puffins at the Hermaness colony seem not to have reared any young at all in 1987 and 1988. In all three cases, the cause was lack of food: because the local sandeel fishery was so intensive, so few of the fish remained in the sea that the chicks simply starved to death.

There is no doubt that the sandeel population has fallen to a critical level. The wider statistics show stocks of other industrial and commercial species, without exaggeration, to be in a perilous state. The colonies of seabirds that live and breed around the sea depend entirely on the fish for their existence – and they will survive only if the fish are protected by sufficiently strong measures. But there need be no catastrophe. Pragmatism, common sense and a common purpose could balance ecological with commercial interests.

While stocks last
Fulmars and seagulls benefit from the offal and whole fish discarded by fishing vessels. Perhaps as many as 2.5 million birds in the North Sea are supported by this temporary bonanza.

Winged casualty
Damaged fishing nets thrown into the sea prove fatal for birds that become entangled in them. The use of persistent synthetic fibres exacerbates the problem.

SEALS, DOLPHINS AND FISHERIES

Large numbers of fish attract not only fishermen, but seals and dolphins too, both in search of an easy meal. But the competition with the fisheries can be lethal for the marine mammals, which may accidentally get caught in fishing gear and drown before the nets are brought up. This is bad news for the fishermen as well, because as the creatures make their desperate attempts to escape they often damage the nets.

Reports suggest that the problem is significant, although there have been few systematic studies. The most detailed was a partial survey of Danish demersal fishing vessels in the North Sea, Skagerrak and Kattegat carried out in 1983. It recorded 149 porpoises killed, an average of one per trip. From this it was estimated that the Danish fleet alone was accidentally catching some 3,000 porpoises a year.

Seals also face a problem with gill nets, large walls of net that are usually anchored in position in coastal waters. The mesh used is so fine that the nets are difficult to detect, and the seals swim straight into them.

Both seals and dolphins get entangled in damaged sections of nets that fishermen cut out and throw overboard while making repairs. This problem can be easily prevented if the damaged sections are brought back to land.

It may be possible to reduce the so-called bycatches of marine mammals by incorporating 'trapdoors' in fishing nets to allow the animals to escape, or by using acoustic devices that deter them from entering. Setting certain parts of the sea off-limits for fishing, to allow fish stocks to reach higher levels, might also be beneficial to marine mammals – as well as to the whole ecosystem.

FARMING MUSSELS IN THE WADDEN SEA

A huge mussel farming industry has grown up in the Dutch Wadden Sea, from an annual yield of 750 tonnes before 1950 to anywhere between 30,000 and 130,000 tonnes in the 1980s – the fluctuation depends on the state of the market and the number of young mussels that are available each year. This is now by far the largest shellfish industry in the North Sea.

Mussels grow naturally in beds on the mud flats throughout the Wadden Sea. Farmers remove very young mussels from their seed beds, placing them in more easily accessible areas where currents ensure a good supply of food, and where they can be easily protected from predators, until they are large enough to be harvested.

This is a terribly wasteful exercise: about the same weight of 'seed' mussels is required to produce the fewer but larger mussels that eventually reach the market place, because so many die in the process of being moved to the new site.

The inefficiency of the industry affects wildlife in the region that depends upon mussels for food. A single eider duck needs about 2.5 kilograms of mussels a day, and eiders probably remove 55,000 tonnes of mussels each year from the Dutch Wadden Sea alone. This represents one third of all the food eaten by birds in this part of the sea. Shore crabs take a similar amount of mussels, and starfish also feed upon them. Fishing for seed mussels causes serious damage to the natural mussel beds. Many birds and other marine life are drawn to the beds immediately after the fishing because the dead and dying mussels left there provide an instant source of food. Once these have been eaten, the area supports far less life than it would if the beds had been left alone.

One way to reduce the impact of mussel farming on the natural environment is for farmers to take fewer seed mussels and settle them at lower densities on the nursery beds. This can yield the same weight of mussels and individuals will be bigger. Best of all would be to cultivate mussels completely independently of the wild population. The beds could also be sited where the mussels would take up the excessive amounts of nutrients that come from farming and sewage works before they enter the wider Wadden Sea and cause over-fertilization. However, for the moment the Dutch mussel fishery serves only to illustrate the conflicts between fishing and the natural world.

Mussel beds
In the German and Danish Wadden Sea mussel fishing is still based on natural beds, and remains on a fairly small scale.

RESTORING THE SEA

By 1983 the North Sea states had recognized the dangers inherent in overfishing, and they agreed to use the scientific models to set target catches (total allowable catches, or TACs) for each species of fish. The models require detailed information about the life cycle of each species; when very little is known, a 'precautionary TAC' is set, which gives the fish the benefit of the doubt. But not enough is known about any species to define a maximum possible catch consistent with its survival and, in practice, the setting of TACs has exposed the biological, economic and social shortcomings of fisheries management.

Some fisheries biologists now recommend that the TACs of most fish stocks be halved; or even that fishing be prohibited in parts of the North Sea. Some argue that fisheries management should, for the present, simply aim to maintain the spawning population above a level that has proved consistently able to ensure that enough fish reach adulthood. Both suggestions are more pragmatic, and more likely to succeed, than the use of elaborate but inaccurate models.

Economic issues seriously undermine the TACs. Nobody agrees how the limited catch should be shared among the various fishing fleets. Just setting the figure for the total number of fish that may be caught results in a glut of fish on the market as each fleet catches as many fish as it can in the shortest possible time. More rational approaches include giving quotas to individual vessels, for instance on a weekly basis; and issuing only a limited number of fishing licences. Even then, a catch of large cod fetches the highest price, so in order to make the most of their quota, fishermen must throw back the smaller fish, most of which are by now dead or dying. In some cases more than half the catch will be rejected. In addition, when the fish are landed, there is virtually no system of enforcement to verify the size or even the species of catches.

The long-term goal of reducing the capacity of the European fishing fleet to match the quantity of fish available is intended to settle the matter by 2002. That time cannot come soon enough for the North Sea; but for the fishing communities it could be devastating, unless great care is taken.

End of the line
A small part of the 200,000 tonnes of haddock caught annually in the North Sea. The catch has fallen from 500,000 tonnes or more since the 1970s, and now virtually all the fish caught are so young that they have barely had time to breed.

Looking at broader issues

Fisheries management cannot be successful unless it deals with the social as well as the biological implications. All around the North Sea, fishing communities feel their livelihoods threatened. Fishermen consider that they are hostages to a decision-making process that is largely outside their control. They are asked to dump fish caught over quota. They are confined to port while ships from far away remove the protected stock, sometimes as a bycatch that is then dumped. The fishermen of the different nations use different equipment and harvest different fish intended for different markets – all of which make the problem of overfishing difficult to deal with. Many feel locked into a system that is careering towards the commercial extinction of fish stocks; a mad gambling game in which they either have to raise their stake to buy ever more expensive equipment to stay in the game, or get out, often with considerable losses.

The social risks inherent in successful fisheries management are at least temporary. Not taking immediate action risks the collapse of the North Sea fishing industry much sooner. But if the fishing pressure is relaxed quickly and completely, it is hoped that many species of fish will recover relatively fast. Although the spare capacity of the fishing fleets will be removed by 2002, fishing communities will recover as the fishing once again becomes profitable; and in the meantime the economic costs of shielding them should not be excessive.

Until now, fisheries policy has, by and large, been to react to events. In future it should fix firm objectives and stick to them – then it would benefit the species that live in the North Sea and the people around its shores.

Small operation, large problems
A small-scale fishing operation, setting fixed nets. The problems of overfishing affect fishermen just as much as the environment. Decisions are often taken by the fisheries management without reference to fishing communities, and show little understanding of how fishing fits into the local economy.

NUTRIENTS
TOO MUCH OF A GOOD THING

IN EARLY MAY 1988 one of the most massive blooms of algae the world has ever seen occurred in the waters of the Skagerrak and Kattegat and along the southern coast of Norway. In a single day, a thick soup of single-celled plants, of the species *Chrysochromulina polylepis*, with tens of millions of cells per litre of water, spread over more than 50 kilometres. Rapidly, it spilled over the surface of the water at the Swedish end of the Skagerrak and began spreading westwards. By 14 May it had reached Arendal, on the Norwegian coast of the Skagerrak; by the 20th it had pushed forward another 120 kilometres and turned the southern tip of Norway. Five days later it had reached Stavanger, 120 kilometres up the coast.

As the algae spread through the water, they released a lethal poison, which had a devastating effect on marine life. By the time the bloom died away at the beginning of June, the seabed was littered with dead lobsters and crabs. Strong-swimming fish, such as adult cod, ling and herring, had been able to flee, but young cod and whiting, too small to escape, died in their thousands. Others, such as wrasse and gobies, were forced to seek refuge beneath rocks or in crevices on the seabed, and there they perished. In coastal waters, huge numbers of organisms – delicate red seaweeds, cat worms, whelks, topshells, periwinkles, limpets, starfish, sea-urchins – were damaged or destroyed. Around 600 tonnes of farmed fish in Norway and Sweden were killed, and many more would have suffered had not fish farmers taken the unprecedented step of towing the huge pens into the fresher and safer waters of the fjords.

No one could recall such a massive bloom, stretching over 1,000 kilometres of coastline from north of Stavanger through to the Danish archipelago in the south, nor one that had had such appalling consequences. Nevertheless, some people believe that it was merely an extreme example of a perfectly natural phenomenon.

Rescue operation
When the bloom of *Chrysochromulina polylepis* (above: the alga shown in an electron micrograph) struck Norwegian waters in 1988, pens of farmed fish had to be towed to safety (below). The pens could be moved only very slowly, to make sure that they did not break up, and that the fish swimming along inside them did not become exhausted.

Natural blooms

Each year as winter gives way to spring, increasing amounts of daylight bring to life the phytoplankton lying dormant in the North Sea. As the surface of the water warms up, it becomes less dense than the colder water below, and becomes distinct from it. Trapped in this discrete layer, the phytoplankton are bathed in the sunlight they need in order to reproduce, and their numbers multiply rapidly.

But soon the zooplankton that feed on the phytoplankton stir into life and their numbers begin to build up, too. They eat away at the phytoplankton, which are themselves gradually using up the supply of nutrients in the surface layer. In little more than a week of the zooplankton stirring, there will be barely enough food to support them, and their population, in turn, will fall drastically. Then, throughout the summer, zooplankton and phytoplankton bump along in much smaller numbers. But the first gales of autumn push pulses of nutrient-rich water from the bottom into the surface layer and there may still be enough light for the phytoplankton to bloom a second time. If they do, the zooplankton also multiply briefly, before the layers of water break down with the coming of winter.

This is the natural cycle of spring and autumn blooms that typically occurs in the northern and central North Sea. In the shallower south, the turbulence caused by strong tides keeps the water mixed all year and usually prevents a separate surface layer forming. Here, although there may be an initial surge of growth in spring, the phytoplankton enjoy pulses of growth throughout the seasons.

Unnatural aftermath
In the wake of the *Chrysochromulina* bloom, starfish enjoyed a massive population boom. The toxic algae had killed the seaweeds and animals living on the rocks in shallow waters; vast numbers of mussels had quickly settled on the empty rocks, providing a superabundant supply of food for the starfish – and stimulating a massive increase in their numbers.

A massive natural bloom
A satellite spies a huge natural algal bloom, more than 500 kilometres long and 150 kilometres wide, in the North Sea between Scotland and Norway in July 1982. The bloom shows as milky white in the picture; clouds and land both appear black. Parts of the coast of mainland Scotland, Orkney, Shetland and Norway are all visible. The bloom is that of a type of algae known as coccolithophore, almost certainly *Emiliania huxleyi* – a species that regularly blooms in the spring in northern waters.

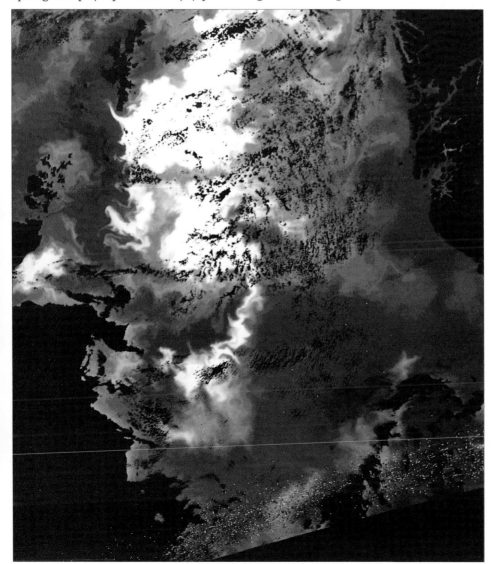

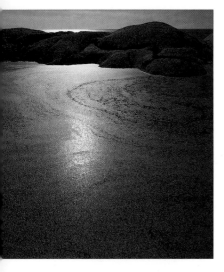

A sea thick with algae
The waters of the Skagerrak suffered an increasing number of algal blooms in the 1980s. The area accumulates nutrients carried there from the southern coasts of the North Sea, as well as from local farmland.

Increasing blooms

So, blooms of phytoplankton are perfectly normal. But over the past few decades they have been occurring abnormally often, growing ever larger in size, and having a correspondingly greater effect on marine life. Coastal waters especially are suffering the highest number of blooms – and it is in those regions that the nutrients on which the phytoplankton feed are at unprecedently high levels.

Phosphorus and nitrogen are the major nutrients phytoplankton need. Indeed, all forms of life depend on them, and they are present in roughly similar proportions in both the marine and the terrestrial world. When we upset the balance of nature, by adding excessive amounts of the nutrients in a short time, eutrophication – overfertilization – takes place and the environment's problems begin. Agriculture, the system of sewage disposal, and the burning of fossil fuels all, directly or indirectly, contribute nitrogen and phosphorus to the North Sea.

If there were no nutrient pollution, each year around 273,000 tonnes of nitrogen would enter the North Sea via the rivers and 1,724,000 tonnes would come from the Atlantic. By 1980 the actual figure had reached around 2,724,000–2,824,000 tonnes. A further 220,000–400,000 tonnes came from the atmosphere directly into the North Sea. Similarly, 18,000 tonnes of phosphorus should naturally come from rivers each year and 232,000 tonnes from adjacent ocean regions. But, by 1980, the figure had reached 359,000 tonnes, with an extra 10,000 tonnes from the atmosphere.

As a general rule it is the nitrogen in the sea that controls the phytoplankton population: when more nitrogen is added, more phytoplankton will grow. If a bloom occurs, an enormous amount of dead and dying algae rains down on the seabed, and the micro-organisms that feed on it use up great quantities of oxygen in the process. If the currents in the water are slight, very little oxygen will be replaced, and most of the creatures that live in the benthic zone at the bottom of the sea will die of suffocation, leaving the seabed a barren and desolate place. The devastation is potentially greater still if the species of algae is one that, like *Chrysochromulina polylepis*, produces toxins when it blooms.

Gathering the evidence

In recent years, areas such as the Southern Bight, the German Bight, the Skagerrak and the Kattegat have shown a dramatic increase in nitrogen. Between 1962 and 1984, there was a fourfold increase of nitrogen in the German Bight – and a fourfold increase in phytoplankton. Further down the coast, an increase in nitrogen in the 30-kilometre-wide band of water off the Netherlands meant that the annual crop of phytoplankton virtually doubled between 1930 and 1980.

The link between the amount of nutrients, particularly nitrogen, and phytoplankton growth shows up clearly all along the coast of continental Europe. Here, there is a flow of water running north from France to Denmark into which major rivers discharge, one after the other, and in which the levels of nutrients progressively increase. This is matched by an increase in both the annual crop of phytoplankton and the number of exceptional blooms.

Overall, Danish waters have suffered most from algal blooms. Considerable areas of the German Bight have also been affected. But no country around the North Sea has been left untouched. Algal blooms and deoxygenation are also known to occur off the shores of the southern North Sea, in the estuaries of the Thames, Tyne and Tees in England, and in the Norwegian fjords.

While several species of algae produce colourful effects in the water, none can match the species *Phaeocystis* for drama. Its tiny cells are embedded in gelatin, which gets whipped up by the waves and forms huge drifts of foaming slime on beaches. It was the sight of this that, in the 1980s, first aroused public concern about the damage caused to the North Sea's ecosystem by overfertilization.

MORE IS LESS
There is a twist in the tale of algal blooms in the North Sea: there is actually less of one kind of phytoplankton – the diatoms – than there was 40 years ago, and nutrients may be responsible. Diatoms are unusual in that they build their cell walls using relatively scarce silica, which they absorb from the water along with the other essential plant nutrients.
The high levels of nitrogen and phosphorus in the rivers may encourage the growth of freshwater diatoms; these use up large amounts of silica, stopping it even reaching the sea – and the marine diatoms perish without it.

DEADLY BLOOMS

The bloom of *Chrysochromulina polylepis* in the spring of 1988 so swiftly devastated the life in the North Sea because it releases a potent poison that disrupts the nervous system. Why algae should produce toxins – what purpose they serve – is still a mystery. They could be waste products that just happen to be poisonous, or they may be designed to kill the zooplankton that feed on the algae. Shellfish such as mussels seem to be immune to the poison, which they store temporarily in their flesh; if people (or birds or fish) eat them they may become seriously ill – some may even die (this is why the harvesting of shellfish in waters prone to algal blooms may be banned at certain times of the year).

Some species of toxic phytoplankton, such as *Gonyaulax* or *Gyrodinium*, bloom so profusely that they reach densities of 20–30 million cells per litre of water; then, the sea takes on a distinct hue – it may turn red or yellow or brown, depending on the species of algae, although the phenomenon is always known as a 'red tide'.

Red tides have been reported since biblical times, which some government scientists hold up as evidence that nutrients discharged by modern society cannot be responsible for them. But this is not really the issue. What has to be established is whether pollution by nutrients – or by any other substance – may increase the chances of toxic blooms occurring.

The Japanese connection

The most persuasive evidence for the link between pollution and red tides comes from the Seto Sea in Japan. There, in the 1960s and 1970s, increasing coastal pollution was matched by a growing frequency of red tides and the loss of an increasing number of commercial fish. When the authorities took steps to control the pollution, the number of red tides declined.

Red tides in the semi-enclosed Oslofjord in southwest Norway have also been associated with an overload of nutrients from coastal discharges in recent years, and toxic algal blooms in this region have increased dramatically during the last decade. In particular, the species *Gyrodinium aureolum* has left unusually large numbers of dead fish in its wake; and *Gonyaulax excavata* and species of *Dinophysis* now make the blue mussel unfit for human consumption for most of the year.

When the bloom is over
Fish killed by the nerve poisons of toxic algae may themselves be poisonous to both marine mammals – seals and dolphins – and people.

Red for danger
The spring phytoplankton bloom at its peak in mid May 1980. In this false-colour satellite image, the blue areas are those where there is little phytoplankton; and the red and white areas, near the coast, are those with very high levels.

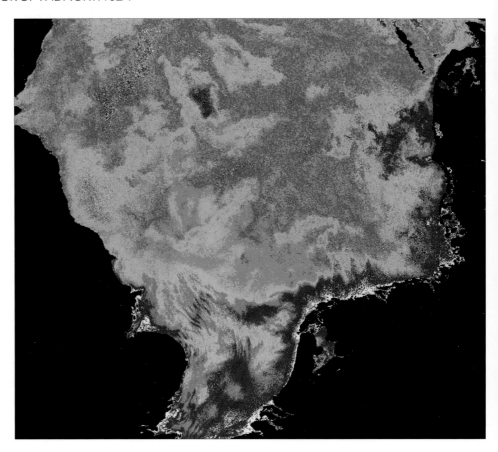

A plague of foam
The alga *Phaeocystis* caused the foam and slime that plagued the beaches of northeastern Europe during the 1980s. Although *Phaeocystis* has been present in the North Sea for many years, it has bloomed much more frequently in the past decade – which is almost certainly due to the increase in the amount of nutrients entering the sea.

THE SOURCE OF THE PROBLEMS – AND SOLUTIONS

There is no mystery about the cause of the overfertilization of the North Sea. The overload comes from three major sources, two of which have developed enormously in the past 40 years. The first is intensive farming; the second is the dramatic increase in the burning of fossil fuels; and the third, much more longstanding, is sewage effluent and the means of its disposal.

Nutrients from agriculture

About half of the nitrogen and about a quarter of the phosphorus (amounts vary slightly from country to country) in the North Sea comes from intensive livestock units and from arable farming. In traditional mixed farming, animal manure – a mixture of dung and straw – is returned to the fields to fertilize them and improve the texture of the soil; few of the nutrients it contains are lost to the rivers. But in intensive animal farming, dung is a waste product, a problem to be disposed of at the lowest possible cost.

Most of the waste is simply reduced to a slurry and applied to fields as fertilizer. But because it is a liquid it tends to clog up the soil, rather than condition it as manure does, or it is quickly washed out into the rivers or the water table, taking the nutrients with it. Such an enormous quantity of slurry is produced – anywhere between two and ten times the amount of human sewage – that every year it is more difficult to find farmland with spare capacity to absorb it. If more is applied than the land and crops can use, the nitrogen in the slurry evaporates as ammonia into the atmosphere, or is leached away into rivers and, eventually, into the sea. In order to reduce the amount of nitrogen run-off from the land, slurry should be applied only at the start of the growing season, when plants will most readily take it up; and if it is applied to bare soil, it should be ploughed in, rather than simply spread on the surface.

Alternatives have been proposed – such as mixing it with straw to try to produce a material closer in its properties to manure; fermenting the slurry to produce methane gas, which could then be used as a source of power for the farm; setting up small treatment plants in the livestock units themselves; or, ultimately, expanding the rural sewage system so that it can cope with removal of the slurry itself if nothing else works. But all these potential remedies have their difficulties and involve an extra expense that comparatively few producers seem prepared to meet. It may well be that if solutions cannot be agreed, the intensive methods of production themselves need to be questioned.

Excessive use of chemical fertilizer, which contains phosphorus and nitrogen, causes similar problems: nitrogen is rapidly leached from the soil, leading to high levels of nitrate in groundwater; and although phosphorus is not so easily leached, because it tends to become bound chemically to the soil, some still may be washed away by rain to enter the rivers flowing into the sea. Practices such as growing just one kind of crop and 'set-aside' schemes to reduce agricultural surpluses result in land lying fallow in winter: with no plants to take up the nitrogen that has accumulated in the soil, it simply seeps away. The measures needed to prevent this are similar to those that apply to the use of slurry. But most essential is a sound agricultural policy that guarantees that farmers apply only the amount of fertilizer that their crops need, at the time when they need it.

Some North Sea states have made attempts to reduce the quantity of nutrients entering the sea from agriculture. Denmark has introduced the most intensive programme, in the wake of the damaging algal blooms off its coasts. Livestock units are legally required to store slurry and to apply it only at certain times of year, while farmers must ensure that crops occupy a substantial area of the fields throughout winter. By such means the Danes hope to halve the amount of nitrogen they add to the North Sea and to reduce the amount of phosphorus by more than 90 percent; but, because of the quantities involved, it is unlikely that they will achieve their target date of 1993. Other countries, too, have introduced, or are considering, schemes that incorporate some of these measures.

New farming, new problems
Pigs confined to pens indoors (above) and huge fields of barley and rape (below) are typical images of today's intensive farms. Modern farming methods lead to enormous amounts of slurry and fertilizer being carried from the soil into the sea.

Nutrients from sewage

About one quarter of the nitrogen and one half of the phosphorus that we put into the North Sea comes from human effluent, which is discharged into the communal sewage system and purified to some extent before joining the natural water cycle. The degree of purification depends on the sewage plant. Some are equipped to give only the most basic level of treatment; other, more sophisticated installations treat the sewage in several different stages before discharging it.

In the primary, mechanical, treatment sewage is simply screened to remove large solids, then reduced to a slurry and allowed to settle. The liquid that results from this process is cleaner, but it still contains most of the nutrients. It also contains other organic impurities, and micro-organisms that feed on these take large amounts of oxygen from the water into which the sewage is discharged.

If the sewage is given secondary or biological treatment, bacteria are used to break down the organic matter. This is speeded up by trickling primary sludge over small rock chippings with a large surface area, or by constantly stirring the sludge, either mechanically or by air. This sewage will now remove far less oxygen from the water into which it is discharged, but most of the nitrogen and phosphorus still remains.

It is only in a third stage of treatment that most of the nitrogen and phosphorus may be removed. Adding certain chemicals to the sewage causes the precipitation of phosphorus compounds; these can, in theory, be recycled. Various other processes attempt to remove as much nitrogen as possible. Air may be blown through the effluent, but this simply puts ammonia (a compound of nitrogen) into the atmosphere. The nitrogen will eventually fall to the ground again in rain, so this treatment merely distributes the problem over a wider area. Another approach is to grow algae in the water to absorb nitrogen compounds, and then strain the algae from the sewage before it is discharged. In a very few installations, bacteria may be used to remove the nitrogen, but this involves a complex set of operations and there are at present very few plants with these facilities in Europe.

Choked with life
A mass of green algae – encouraged by fertilizer flushed from the surrounding land – chokes a drainage canal, displacing the rich assortment of plants and animals that once flourished there.

The sludge that results from the various treatments is allowed to dry out, then is usually burned or dumped in a quarry or landfill site. In Belgium, however, even major cities such as Brussels simply discharge sewage effluent into rivers, and from there it finds its way into the North Sea. The United Kingdom dumps nearly a quarter of its sewage sludge in the North Sea, after giving it only a basic level of treatment, primarily to reduce the volume of the waste and its toxicity. Many coastal towns in the United Kingdom discharge sewage from a pipe that extends, at most, just a few kilometres offshore. In Holland, too, 11 percent of sewage is given only primary treatment before it is discharged from a pipe extending 10 kilometres into the North Sea from The Hague.

In some areas, the sludge is used in agriculture, giving soil and crops the benefit of at least some of the nutrients. But, like animal sludge, it should be applied only at certain times of year to reduce as far as possible the quantity of nutrients being washed from the soil into rivers. For this to be effective on a large scale, the cost of transporting sewage sludge from the treatment works, usually in urban areas, would need to be reduced substantially, especially given the considerable quantities of animal sludge that are more readily available. In any case, the practice of using sewage sludge in agriculture has a significant drawback: the sludge may be contaminated by domestic chemical products and industrial waste, containing toxic chemicals and heavy metals, which is also discharged into the sewage system. If contaminated sludge is applied regularly, the toxins build up in the soil and will affect any crops grown in it.

Nutrients from the atmosphere

About a quarter of the nitrogen overload in the North Sea comes directly from the atmosphere, but an additional major contribution falls in rain over land and is hidden within the statistics for inputs from rivers. Nitrogen gets into the atmosphere mainly in two ways: from the exhausts of vehicles with petrol or diesel engines, and from power stations burning fossil fuels (gas, coal and oil).

In both cases, the process of combustion means that oxygen and nitrogen already present in the atmosphere are fused at high temperature into nitrous oxides. The higher the temperature, the more nitrous oxides are formed.

Trailing the rest of Europe
The government of the United Kingdom has said that it intends to continue dumping sewage sludge at sea until the late 1990s. All other North Sea states stopped in the early 1980s.

Sewage agitation
The removal of nitrogen from sewage effluent using a series of bacterial treatments, such as at this plant in Sweden, is still a rare practice in Europe.

Cold weather tells all
Frost and snow reveal the scale of the emissions from vehicle exhausts. The nitrous oxides in the exhaust gases (as well as other pollutants such as lead) contribute to pollution entering the sea.

Road traffic Vehicle exhausts are responsible for about half the nitrous oxides that pollute the atmosphere. Although various regulations to reduce this figure have been introduced across Europe, the effects have been outstripped by the increasing number of cars. For example, although West Germany introduced measures to reduce emissions from vehicles from 1,000,000 to 750,000 tonnes between 1985 and 1988, the figure actually rose during this period, to 1,100,000 tonnes, because more vehicles came into use.

Britain's transport policy reflects this contradiction. It predicts that road traffic will increase by between 83 and 142 percent in the next 30 years – but the increase is not taken into account in its proposals to reduce vehicle emissions. Different European countries have put forward varying standards for emissions, and have differing expectations for traffic increase, but the pattern is broadly the same from Switzerland to Sweden.

A European Commission Directive passed in 1989 requires that, from 1992, all new cars are fitted with a three-way catalytic converter, which reduces carbon monoxide, unburned hydrocarbons and nitrous oxides produced from petrol engines. Fitting a three-way catalytic converter to an engine can reduce its output of nitrous oxides by about 70 per cent. But converters work effectively only on new vehicles, so it will take some time to realize the full benefits of this system of cleaning exhaust gases; and more research is needed into cleaning gases from diesel engines because, although they produce only 60 percent of the nitrogen compounds of a petrol-driven car fitted with a three-way converter, they do produce highly toxic polyaromatic hydrocarbons.

Another way to lower emissions is to modify the combustion process itself. Ordinary car engines produce waste hydrocarbons and carbon monoxide because of incomplete combustion. The motor industry has for many years been trying to produce a 'lean burn' engine, which increases the ratio of air to fuel in such a way that combustion is both faster and more complete; but the technological problems are proving intractable.

However, every motorist could help to reduce this pollution immediately – simply by slowing down. A car driven at a constant speed of 100 km/h may give out four times as much nitrous oxide as one driven at 60 km/h. This could be the most effective short-term way to lower nitrogen oxides – and it costs nothing.

Power production Emissions from power stations – particularly those fired by coal – presently account for most of the remaining nitrous oxides in the atmosphere. The best approach here would be to switch to a fuel such as natural gas, which is not such a pollutant; or to reduce the emissions from coal by lowering the temperature at which the fuel burns. As an interim measure, power station chimneys could use the catalytic reduction process to remove nitrous oxides from the flue gases before they are released into the atmosphere.

Another way to lower nitrogen emissions is called fluidized bed combustion. Coal is crushed into a fine powder and fluidized by suspending the particles in a strong current of rising air. The result is rather as if the powder were in a boiling liquid. Unlike conventional systems, the boiler tubes that carry away the steam (to drive the turbines generating the electricity) can be placed in the middle of the burning fuel. Heat can be collected more efficiently, so the system can operate at only half the temperature of present power stations.

An alternative is to produce clean-burning gas from coal and burn this in a gas-fired power station. Nitrous oxides are reduced by adding water vapour to the gas, lowering the temperature at which it burns. Again, this results in lower emissions compared with those from contemporary coal-fired power stations, and at levels similar to those generated by natural gas.

Yet another way is to put more effort into generating electricity from renewable energy sources – using wind, wave and geothermal power. But the simplest and quickest way to reduce emissions is still by conserving energy.

A mountain of coal
The combustion of fossil fuel, and of coal in particular, is a major source of nitrous oxides.

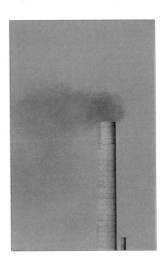

Smoke signal
These distinctive brown fumes signal nitrous oxide pollution from a chemical industry.

POLLUTION
THE INDUSTRIAL OVERLOAD

THE NORTH SEA is the ultimate sink for vast quantities of pollutants. By the end of the 1980s, at least 23,600 tonnes of zinc, 6,800 tonnes of lead, 4,400 tonnes of copper, 4,200 tonnes of chromium, 1,450 tonnes of nickel, 820 tonnes of arsenic, 150 tonnes of cadmium, and 50 tonnes of mercury were being added to the North Sea each year. In the coastal waters around the North Sea, where the metals become trapped by the inshore currents, zinc concentrations were 20–50 times higher than natural levels in the North Atlantic; lead was 15 times higher; nickel was 3.5–5 times higher; copper 3–6 times higher; cadmium 3–10 times higher; and mercury 10 times higher.

The cocktail is made even more potent by the addition of literally thousands of man-made chemicals each year. Unlike heavy metals, which occur naturally in very small quantities, such substances as halogenated hydrocarbons (HHCs) are totally artificial and have absolutely no place in nature. Yet no systematic attempt has been made to calculate how much or how many of these chemicals are being discharged into the North Sea. Only a handful of them are monitored – synthetics like the pesticide lindane, polychlorinated biphenyls (PCBs), and the 'drin' family of aldrin, dieldrin and endrin – even though we know that minute quantities of many others have horrific effects on the living organisms that unwittingly absorb them. Humans and animals alike find it difficult or impossible to break them down and eliminate them from their bodies.

Blue tide
A discharge of synthetic chemicals from a biochemical plant pouring into the Ouse in England, 20 kilometres from the North Sea. Even in the 1990s, pollution of the rivers running into the North Sea continues on a massive scale.

So does the North Sea. Virtually surrounded by intensive industry, and with a strictly limited ability to dilute the destructive agents poured into it, the North Sea has become one of the most heavily polluted marine areas in the world.

The problem of bioaccumulation

Life is a miracle of the accumulation of matter from very low concentrations of thousands of different minerals that occur naturally in the environment. In the sea, salt is the most abundant mineral, yet it makes up just one third of one percent of the total composition. Other chemicals essential to life are present in much smaller quantities. Carbon dioxide, for example, the raw material of all living matter, is present at a concentration of just 28 millionths of a gram per litre of water. Natural levels of nitrogen and phosphorus, the important ingredients in protein and DNA respectively, can be measured only in millionths of a gram per litre – and in spring the levels fall rapidly, to only billionths of a gram, as the nutrients are absorbed by increasing amounts of phytoplankton. Some essential micronutrients are also measured in billionths of a gram per litre. Natural concentrations of the heavy metals range from 5 billionths of a gram per litre for zinc, down to 30 trillionths of a gram for mercury and lead. (There are, of course, no natural levels for synthetic chemicals.)

The difficulty for the natural world begins when these concentrations are increased – by even tiny amounts. Small changes in salinity, for example, are lethal to many marine species; tiny alterations in levels of nutrients can have a major effect on the abundance of different phytoplankton. When creatures absorb HHCs and heavy metals the problem is exacerbated because they often cannot excrete the toxins. Some pollutants are not readily soluble in water, but they do dissolve in fat and are stored in fatty tissues. And there they accumulate, small quantities building up to high levels in a process known as bioaccumulation. The situation gets worse the higher one moves up the food chain: when one creature eats another it also absorbs the pollutants that that animal has eaten, and the toxins become ever more concentrated. As a result, the levels of pollutants in living organisms can be far greater than those found in the environment.

Too weak to resist
Heavy metals and synthetic substances accumulate in living organisms through the food web, reaching their highest concentrations in marine mammals, such as the harbour porpoise. The effects of the chemicals are to reduce the creatures' ability to reproduce and to impair their ability to fight disease.

Red carpet, dirt cliff
Pollutants do not have to be toxic to have lethal effects. Iron and aluminium discharged from an aluminium smelter in Germany (left) and powdered rock from a coal mine on the English coast (above) blanket both land and water, smothering all life below.

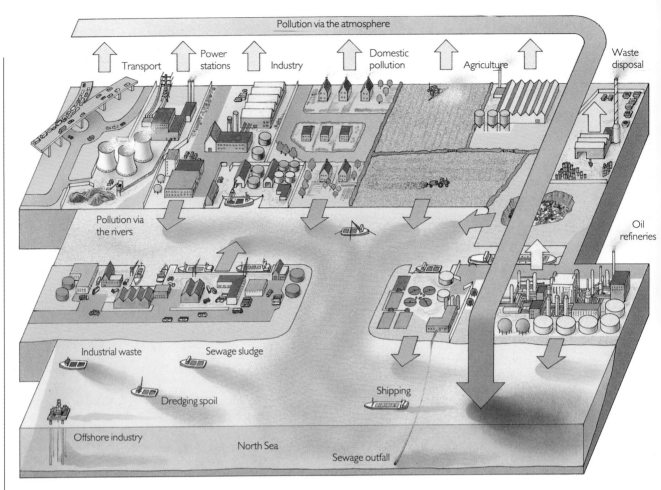

HEAVY METALS

Because heavy metals occur in nature, it has been assumed that the environment has the ability to assimilate them – a way of absorbing the concentrations created by human industry back into the natural cycle. The assumption developed from the belief that nothing humans could do would upset the balance of nature – but that belief was far too modest.

Many metals are among the building blocks of life. The chlorophyll in plants, which makes photosynthesis possible, contains magnesium; blood is based on iron. Zinc and copper are constituents of enzymes, proteins and vitamins. However, most heavy metals are needed only in minuscule amounts; if they are present in living organisms at anything higher than trace levels, they become toxic. This is especially true of mercury, cadmium and lead, whose exact biological function – if any – is still not known. Mercury damages the central nervous system and impairs reproduction. Cadmium attacks human kidneys and causes skeletal deformities in fish. Lead is particularly dangerous to children, and it can cause extreme mental and behavioural disturbances in many animals.

Whatever their biological role, heavy metals are major contributors to the material quality of life that most people take for granted. Cadmium is used in the manufacture of plastics, in rechargeable batteries, to process other metals, and is released during the production of fertilizer. Copper is used in electrical equipment. Chromium and nickel are essential to alloys and electroplating. Lead is still used in petrol, while mercury is vital to the production of chlorine and caustic soda – two of the most basic raw materials of the chemical industry. The toxic nature of some compounds, such as those based on tin, has even been deliberately exploited, to produce anti-fouling paints used on boats, on docks and quays, and on fishing tackle to prevent the growth of algae and animals. There is no doubt about their efficiency: they have proved acutely toxic to many forms of marine life in coastal waters.

The routes of pollution
The North Sea ultimately receives much of the pollution that is generated in northwest Europe. Pollution may be directly released into the sea, but it also comes via the rivers and from the atmosphere. Pollution comes from the food we eat, the fuel we burn, the journeys we take, and the products we make, use and discard.

POLLUTION

Out of sight...
A municipal rubbish tip (left) – an irresistible lure to flocks of scavenging herring gulls – is a long-term source of pollution. Heavy metals and synthetic substances from domestic and industrial waste leak into the water table.

Fouling the sea
Organometallic compounds were banned as pesticides because they were too dangerous. But their use in anti-fouling paints, which kill marine organisms settling on ships and docks, is still alllowed. The chemicals leach into the water, poisoning the creatures that live in it.

The metals reach the sea in a variety of ways. During the extraction of ores and in manufacturing processes, they are discharged as waste into the atmosphere, or piped into rivers or into the sea. Sewage sludge, contaminated by heavy metals put into the sewage system by industry, is dumped directly from ships. Lead is released into the air from car exhausts, and other heavy metals enter the atmosphere during the incineration of industrial waste. Even when industrial waste is stored in landfill sites, the metals in it frequently leak into the water table and into the river system. Most dumps are unmapped and untreated, the legacy of a century of careless industrialism; as more sites are discovered all across Europe, they create major problems for those who have the task of making them safe.

Levels beyond the limit

Although the intention is to halve the quantity of heavy metals entering the North Sea by 1995, the amount already there is colossal. Any capacity the environment might have had to assimilate waste was exceeded decades ago. The accumulation of metals in the water is made worse by dredging spoil. Over the years, metals have built up in the sediments in the major estuaries, where channels have to be kept clear for shipping. Constant dredging in the Rhine, Scheldt, Thames, Weser and Elbe, and at most major ports throughout the North Sea region, releases exceptionally high concentrations of metals – as much again as those entering the sea from the rivers themselves.

When a heavy metal enters the North Sea, part remains in the water and part becomes bound to biological or mineral particles that fall to the seabed. Here, the seething mass of fauna constantly turns over the sediment and keeps the metal in the surface layer, where it may either be absorbed by animals or re-enter the water. In coastal areas that have high concentrations of metals, the sediment becomes a major source of pollution as the activity of benthic life converts many metals, including cadmium, copper and mercury, to forms that are readily absorbed by living organisms. Mercury, for instance, is converted by microbes in the surface layer from various mineral forms to highly toxic methyl mercury.

When substances like methyl mercury cannot be excreted, they accumulate in bodily tissues and are passed into the food web to the biggest fish, mammals and birds. Cod in the Kattegat carry up to 1.29 milligrams of mercury per kilogram, 30 times the amount in cod from the cleaner waters off the coast of Greenland.

Metals in the air
Heavy metals are discharged into the atmosphere by the metal smelting and processing industry. Ultimately, many of them will find their way into the waters – and the sediments – of the North Sea.

Auks such as black guillemots that feed on fish in coastal waters have higher mercury levels than other auks that feed in offshore waters. Levels of mercury in migrant shorebirds increase during their stay on North Sea estuaries, and higher levels also occur in dolphins and seals. There are parallel stories for other metals.

Single metals have effects that are toxic enough. But when two or more metals work together, their toxicity can be far greater than the sum of the parts. Mercury, lead and zinc together virtually halve the growth of certain single-celled zooplankton in comparison to the effect of each metal in isolation. But the effects of metals can also increase the impact of other pollutants. Worst of all are the problems caused when high levels of metals occur in conjunction with the group of substances called halogenated hydrocarbons.

Chemicals for the sea
Discharges of heavy metals into rivers, such as this flow of chromium from a tannery in northeast England, damage life in both river and sea.

HHCS – THE FUSION OF LIFE AND DEATH

Chemical compounds known as hydrocarbons (which contain atoms of carbon and hydrogen) form the basic material from which all living organisms are built. Hydrocarbons are part of the natural carbon cycle: they break down easily and their carbon components are released into the atmosphere, to be taken up by plants and incorporated once again into living tissue.

One group of artificial carbon compounds, which are much more stable than the natural forms, are created by replacing one or more of a hydrocarbon's hydrogen atoms with atoms of chlorine, fluorine or bromine. These substitute ingredients are called halogens, and the compounds that result are called either organohalogens or halogenated hydrocarbons (HHCs).

Dredging up the past
Such high levels of heavy metals have accumulated in the bottom reaches of the rivers that some countries now store contaminated sludge dredged from shipping channels on land rather than dumping it at sea.

In the late 1940s, when HHCs were first manufactured in large quantities, people believed that their stability would be of great benefit. They are cheap to make (from oil and halogens), they last a very long time, and they are exceptionally versatile. Organochlorine compounds are used in transformer oils, electrical capacitors, hydraulic equipment, and heat-transfer fluids (PCBs); as insulators, flame retardants and plasticizers in some paints, adhesives and plastics (PCTs, PCAs, PBBs, and PCNs); and as a plastic itself (PVC); for dry-cleaning, industrial solvents, refrigerants and aerosols (CFCs); and as pesticides (DDT, lindane, toxaphene and the 'drin' family). But in reaping these benefits, neither manufacturers nor consumers sought to question the long-term disruption that the chemicals, their by-products and their residues can cause to organic, atmospheric and other vital processes. They are a permanent addition to the environment, but they are completely artificial – and often lethal because of it.

HHCs break down only over a very long period of time, in a process that takes tens or hundreds of years. And as they degrade, the products that result are frequently as – or more – toxic and persistent than the original substance. Only rarely can living creatures deal with them – and then they can usually do so only indirectly. Female seals and dolphins, for example, tend to carry lower levels of HHCs than the males, because they pass them to their offspring in their milk.

A sign of the times
The incidence of diseased and deformed fish in the North Sea is worst in those areas that suffer heavy loads of pollution.

The danger of PCBs

The story of PCBs serves to demonstrate the pathways by which HHCs in general escape into the environment, and the damage caused when they do. Because PCBs are known to be so inert physically – that is, they do not react with any other substances – no one suspected that they could cause serious biological damage. But today, PCBs in plankton, invertebrates and fish stand at anywhere between 10,000 and one million times the level in the water, while birds and mammals have accumulated perhaps 100 times more than that. The growth of phytoplankton can be inhibited by concentrations in the water as low as 50 billionths of a gram per litre.

Hope for the future?
Bottlenose dolphins are still present and breeding in the Moray Firth, Scotland, where industry has only recently arrived. The dolphins' continued survival will depend on industry adopting cleaner methods than in the past.

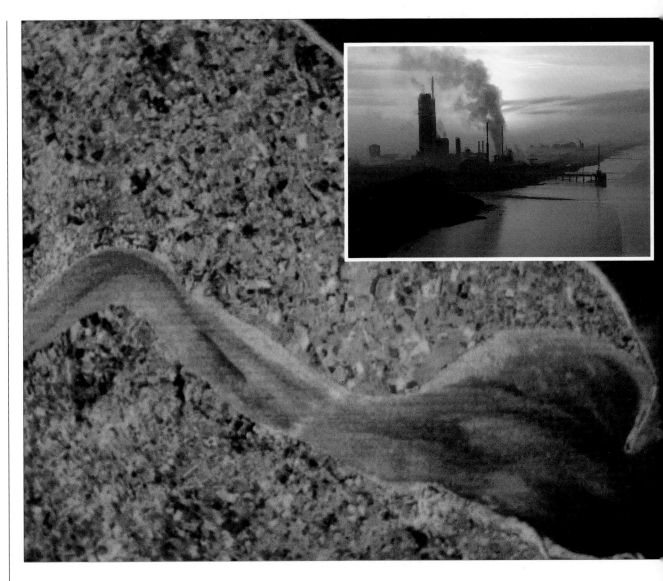

Caught from space
A satellite image of the Humber estuary in England shows atmospheric pollution from industry on the south bank (inset) drifting northwards. Discharges of heavy metals, synthetic substances and oils into the water either settle in the banks of sediment or flow into the North Sea.

West German authorities consider that fish are 'at risk' when ambient concentrations of PCBs are one tenth of those seen to be lethal in the laboratory. Anything between one tenth and one hundredth of lethal levels should give 'serious cause for concern'. For instance, malformation in young fish can be correlated with PCB residues of 0.5 milligram per kilogram of spawn. If these standards are applied to the North Sea, the majority of fish monitored give 'cause for concern'. Throughout the 1970s, the very lowest levels found in oily fish such as herring and mackerel showed them to be 'at risk'; and recent surveys, though less comprehensive, show no change. Indeed, some mackerel have PCB levels as high as 1.06 milligrams per kilogram – double the lowest figure for fish 'at risk'.

In an experiment conducted in the Wadden Sea, seals that were fed on local fish contaminated with PCBs became less fertile (they recovered when their diet returned to healthy fish, from the Atlantic). In Britain, an attempt to reintroduce otters along the East Anglian coast failed, apparently for the same reason. Because PCBs are stored in fat, they cause particular problems in times when food is scarce. As the fat reserves are mobilized, a massive dose of PCBs is released into the creature's system, affecting its ability to fight disease. In 1969, thousands of guillemots in northwestern Europe died in this way.

The North Sea states have agreed to end the manufacture of PCBs during the 1990s, but they will continue to pose a major threat to the environment until safe ways can be found to dispose of them.

A PLAGUE AMONG SEALS

In 1988, a mystery illness devastated the harbour seals that live in the southern regions of the North Sea. The disease arrived in the Kattegat in April, signalled by an unusually high number of seals giving birth prematurely. Within a month seals of all ages were suffering from what appeared to be pneumonia; by June thousands had died. By July, the disease had spread north and south, as far as the west coast of Norway, the Baltic and along the coast of the Wadden Sea. Finally, the epidemic crossed the North Sea itself, reaching the United Kingdom and the Irish Sea in August and September. In many harbour seal colonies, over half of the animals died in the epidemic, making a total of 18,000 deaths.

The agent of the disease proved to be a previously unknown morbillivirus, which has since been named phocine distemper virus. It is related to measles, and to canine distemper, with which it was, at first, confused: both distemper viruses produce similar effects, including suppression of the immune system.

The role of pollution

Some scientists suspect that pollution may have had a role in the severity and extent of the epidemic – but this is a matter of some controversy. It is known that, before the virus took hold, many seals had accumulated high levels of both heavy metals and HHCs in their bodies. Experiments had shown that female harbour seals that were fed fish contaminated with PCBs produced fewer pups than those given less-contaminated fish, and had reduced levels of vitamin A, which helps to resist disease. Another indication that pollution may have increased the severity of the epidemic is that proportionally fewer seals died in Scotland and northern Norway, where the water is less polluted.

On the other hand, it is perhaps only to be expected that a new virus would have serious consequences even for perfectly healthy seals. And levels of pollution in dead seals were not unusually high compared to those of the decade before – but it may be that any contamination above a certain threshold has an adverse effect.

Whatever its role in the 1988 epidemic, pollution undoubtedly affects the North Sea seals. In the most contaminated areas of the sea, seals still carry high levels of pollutants in their bodies, which make them less able to resist any disease, and – because of the effect they have on reproduction – are possibly preventing the population recovering. On this evidence alone, it is clear that emissions of heavy metals, HHCs and other synthetic carbon compounds into the North Sea must be stopped – and quickly.

Death toll
The bodies of dead seals recovered in Germany contained such high levels of pollutants that they were declared 'hazardous waste', requiring special conditions for disposal.

A thousand unknown threats

Many of the chemical products that go under a single name are actually a cluster of similar but not identical chemicals. PCBs include mixtures of up to 100 closely related compounds. Commercial DDT is a smaller family of compounds, but toxaphene, which was introduced as a substitute in many countries when DDT was banned, contains at least 670 compounds.

Altogether, there are thousands of different HHCs entering the environment. Of those whose effects have been studied, many have been shown to be dangerous to marine life. But only a small fraction of the total has been examined because analysis is expensive, or even impossible; and there is no reason to suppose that the others will, by chance, be benign. The only safeguard is precautionary action, and it can be taken at source – on the routes chemicals take to reach the North Sea.

Halogenated waste

HHCs enter the North Sea in the same ways as heavy metals: through the rivers and estuaries, by direct discharge, by dumping and through the atmosphere – by the incineration of industrial waste on land as well as at sea. Few are released intentionally, with one major exception: the HHCs used in pesticides.

While there are increasingly stringent controls on the release of some HHCs, one highly toxic group – the pesticides – continue to be deliberately dispersed. In the Netherlands 19 tonnes of lindane were used in 1974, and 28.5 tonnes in 1985; in the same year West Germany used 120 tonnes. In the United Kingdom the quantity of lindane used on oil seed rape alone more than doubled between 1975 and 1983. More than 20 million tonnes of a lindane–DDT solution have been sprayed over forests in East Germany; and although neither the exact proportions of active ingredients nor the number of years over which the spraying took place are known, it is clear that East Germany makes a significant contribution to the amount of lindane in the North Sea. The concentrations of lindane in the sea doubled between 1981 and 1987. And lindane is just one of many pesticides that are currently in use.

Airborne toxins
Huge amounts of pesticides never reach their intended target (above left), but are carried away in the air, eventually to reach the North Sea and further afield.

Pollution under licence
A mixture of halogenated hydrocarbons and other synthetic chemicals contaminates an English river (above). Extensive discharges may be legally permitted, but even then companies may break the conditions attached to the licence because the chances of getting caught are minimal and the penalties derisory.

Growing awareness of the harm caused by a few HHCs has led laws governing their use to be changed, but has done little to diminish the scale of the damage. DDT was banned in most North Sea states in the 1970s, but its substitute, toxaphene, also proved toxic, forcing its withdrawal in the 1980s. In West Germany over 10,000 tonnes of PCBs have been released into the environment from abandoned mines used as dumps for disused domestic and industrial equipment. In 1983 the UK government estimated that 600 tonnes of PCBs were released every year into landfills from discarded electrical equipment alone.

Finding ways to dispose of these highly toxic substances has proved hugely problematic – and still remains so. For instance, over 1.75 million tonnes of EDC tars – waste created during the manufacture of PVC – were produced in western Europe between 1970 and 1985. At first these were dumped into the North Sea. When this practice was banned, there were attempts to incinerate the waste on land, but many of these were stopped because the large quantities of hydrochloric acid produced corroded the equipment.

Ocean incineration started in 1969, in an area 130 kilometres offshore in the Dutch sector of the North Sea. Throughout the 1980s, some 90,000 tonnes of halogenated waste were burned at this site every year, two thirds of it from West Germany. The process involves burning waste in open-topped chambers, which means that residual gases – sometimes far more toxic than the waste itself – escape freely into the atmosphere and thus into the sea. By the end of the 1980s, halogenated hydrocarbon compounds typical of those resulting from the

Ocean burn
The incineration vessel *Vulcanus II* burns its cargo of chemical waste, considered too hazardous to be disposed of on land – and in the process deposits even more toxic substances in the North Sea. In the late 1980s, incineration at sea released some 5 tonnes of HHCs into the atmosphere every year. The practice is to be prohibited after 1991.

Dioxin purity
Clouds of steam mark the site of a pulp and paper plant in Norway. The pulp and paper industry produces huge amounts of HHCs, including dioxins, when pulp is bleached with chlorine.

combustion process had accumulated in the sediments around the burn site, and in living organisms such as dab and shellfish.

In 1989 Belgium, which until then had received most of the waste for incineration at sea through the port of Antwerp, refused to allow further imports, and West Germany was forced to end ocean incineration. In fact, all countries will end ocean incineration by the end of 1991 at the latest – but this seems set to shift the whole problem back on land.

Dioxins – the dirty dozen

Nearly every form of combustion that involves a chlorinated compound results in a by-product called a dioxin. There are 200 dioxins, of which 12 – known as the 'dirty dozen' – are among the most toxic substances on earth.

The less efficient the combustion process, the more dioxins are formed – and they are created by vehicle engines, iron and steel works, power stations, hospital incinerators, waste incineration and as effluent from a number of other processes. An entire industrial country may produce less than a few kilograms per year from all its sources – but dioxins are so poisonous that scientists calculate a human's 'maximum tolerable weekly intake' to be 35 picograms (35 millionths of a millionth of a gram) per kilogram of body weight, but it may be even lower.

Dioxins disperse particularly well through the atmosphere – which means that they reach the North Sea from Central Europe as well as from coastal regions. The implications for the ecology of the sea are not yet understood; but the implications for cleaning up methods of both industrial production and disposal of industrial waste are obvious.

CHRONIC POLLUTION – THE LONG-TERM EFFECTS

The effects of pollution are easiest to see, and to deal with, when they are acute. When a creature dies from a massive dose of a pollutant, it should be possible to suggest preventative action so that it does not happen again. It is much harder to do so if death is caused by long-term, low level, or chronic, pollution, because the effects can be confusingly subtle.

A Norwegian experiment showed that the growth of phytoplankton in the North Sea can be halted by adding just five millionths of a gram of mercury per litre of water. Other experiments, involving heavy metals and HHCs, reduced the numbers of copepods, which graze on the larger phytoplankton species, and so altered the species composition of the entire phytoplankton community. But a third series of experiments produced the really subtle results.

Between one and fifty millionths of a gram of cadmium per litre of water were added to samples of a typical North Sea plankton community, devastating the numbers of the common comb jelly, which feeds on copepods. In the samples that received low levels of cadmium there were more copepods than there were in the control samples (to which no cadmium had been added), because the number of predators had been reduced. But in the samples given high doses of cadmium, the copepods were affected as well as their predators, and their final numbers matched those of the controls: the increase in copepods due to the absence of comb jellies was exactly balanced by the number killed by the high dose of cadmium. Had this happened in the natural world, anyone simply looking at the copepods – a key element in the food web and therefore likely to be thought a useful barometer of the health of the ecosystem as a whole – would have assumed that all was well.

These experiments demonstrate the difficulties of monitoring pollution. It is impossible to monitor every pollutant and every living organism, just as it is impossible to calculate either the interaction of different pollutants or the variety of complex ways in which a single factor can affect the marine web. It is therefore essential to eliminate or reduce to an absolute minimum the stream of pollutants entering the North Sea – and the best means of achieving this is by adopting the process known as 'clean production'.

Clean production at Landskrona
After adopting the principles of clean production, a metal-working company in Landskrona, Sweden, switched from using solvent-based paints, which created both atmospheric pollution and hazards to human health, to harmless powder-based ones – and made annual savings of at least $400,000. The company was one of a number from 'problem' industries – such as printing and chemicals – that collaborated with Lund University in Sweden to show that it is possible to combine profitability with concern for the environment.

Clean production

Clean production describes a manufacturing process that uses the minimum amounts of raw materials and energy and creates neither toxic waste nor toxic by-products. It also ensures that the products themselves do not create problems for the natural world either during or after use. It shifts the emphasis from waste disposal to waste reduction, but does not include measures that simply reduce the volume of waste after it has been produced, by incineration or concentration, or that merely dilute pollutants before discharging them.

Clean production is actually a series of measures that covers a systematic examination of all waste produced ('waste auditing') and requires better recycling of materials within the manufacturing process. For it to be successful, exchange of information is essential; this is achieved by companies, research institutions and governments working together, pooling knowledge and ideas about raw materials, product design and production techniques. This should go a long way towards ensuring that even the smallest part of every manufacturing process is carried out with the utmost efficiency and economic benefit – and, most importantly, with the least impact on the environment.

Auditing waste

The amount of waste generated by the modern chemical industry is vast, not least because it produces such enormous quantities of chemicals that even substantial

THE PROBLEM WITH NUCLEAR POWER

Around the shores of the North Sea and in its river catchment area, there are 39 nuclear sites: 32 power stations, four research facilities and three plants that reprocess spent fuel. Together, these form the biggest source of radioactive pollution in the North Sea.

In a body of water as large as the North Sea, radioactivity is measured in terabecquerels (TBq) – one terabecquerel signifies that atomic nuclei are emitting one million million radioactive particles per second. In 1988, just over 750 TBq of liquid radioactive effluent were discharged from nuclear power stations into the North Sea and the English Channel, either directly or via the rivers. The discharges have increased by around 70 percent since 1985, mainly because the capacity of nuclear power plants along the rivers Rhine, Meuse, Scheldt and Elbe has increased in recent years.

A powerful risk

Nuclear power stations themselves are not safe: in France, where 70 percent of electricity is generated by nuclear power, the chief inspector of nuclear safety has estimated that the risk of a serious accident in the next 20 years is 'several percent'.

However, the major source of radioactive pollution is the fuel reprocessing plants – at Sellafield and Dounreay in the United Kingdom, and at Cap de la Hague in France. A percentage of the fuel in a reactor has to be replaced each year; when the fuel is removed it is sealed in containers and usually stored for several years in ponds of water to cool down before being sent for reprocessing or to a permanent storage site.

The long-term disposal of radioactive waste is a problem that has never been solved. At present the United Kingdom is proposing to dump waste in containers deep under the sea near its coastal reprocessing sites in the northern North Sea. It is accepted that water will eventually reach the radioactive waste and dissolve it, but it is hoped that the surrounding rocks will absorb the radiation before it escapes into the wider environment. But the likelihood of radiation leaking from the storage sites into the waters of the North Sea and elsewhere is such that other nations are protesting violently at the United Kingdom's plans.

Waste from waste

The reprocessing of fuel produces huge amounts of radioactive liquids, solids and gases, which themselves have to be disposed of in some way. It also involves the transportation of highly radioactive material between power plant and reprocessing centre. And the risk of an accident at any stage is always present. In 1974 at Sellafield, spent fuel rods corroded in their underwater storage tanks, contaminating the sea in turn when the

Deadly trade
Unloading spent nuclear fuel in Cherbourg for reprocessing at Cap de la Hague. Reprocessing plants around the North Sea have made it the international centre for traffic in nuclear waste.

storage water was regularly discharged. It was not until 1978 that filters were installed to trap the radiation leaking from the rods. Even in 1984, radioactive discharges from Sellafield were higher than those from anywhere in Europe.

No research has been done into the long-term effects of radiation on marine life; and there is now perhaps greater uncertainty about what is a 'safe' level of radiation – for humans and wildlife – than at any time in the past. What is clear is that a major accident on the shores of the North Sea must be avoided at all costs – and the best way to ensure this is to end the use of nuclear power.

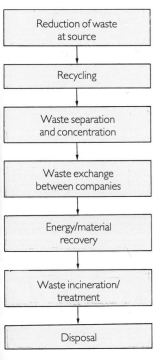

WASTE MINIMIZATION

Minimal waste
The 'waste minimization hierarchy' as defined by the US Environmental Protection Agency. All companies producing more than 100 kilograms of hazardous waste a month are required to have such a programme – and should exhaust the possibilities at each tier of the hierarchy before considering options lower down.

leaks during the production processes may simply not be noticed. One US company producing phenol was leaking 180,000 kilograms of cumene (a synthetic hydrocarbon) every year, costing about $100,000. But this represented just 0.06 percent of the 320 million kilograms of cumene used in the same period. Since the records showed only errors of plus or minus one percent, the leak simply did not show up. Only when an employee noticed the characteristic smell of cumene was the fault, a leaking valve, identified and corrected.

Different plants that use similar types and quantities of hazardous chemicals can generate very different amounts of waste. One plant using 2.5 million kilograms of phenol a year lost 2,000 kilograms (0.08 percent) as waste; while another using 1.6 million kilograms lost a massive 270,000 kilograms (17 percent). Proper schemes to minimize waste can do much to prevent this happening.

In the United States, the Environmental Protection Agency has established a 'waste minimization hierarchy', which requires companies to exhaust the possibilities of using processes that benefit the environment, such as reducing waste at source or recycling material within the production process, before treating waste by any other method.

The examination of each stage of the hierarchy, each step of the ladder, takes the form of a waste audit, which emphasizes the necessity of working systematically through production processes looking for ways to reduce waste. The audit asks:

- What are the waste streams generated by the plant?
- Which wastes are hazardous, and why?
- How much waste is there?
- How much raw material is lost through leaks?
- How efficient is conversion of raw material to product?
- Are unnecessary volumes of waste generated by the mixing of materials, or poor housekeeping?

The answers to these questions provide the information that enables manufacturers to draw up a 'materials balance', which is the heart of a waste audit. It entails carrying out a complete assessment of the weight of both the raw materials that go into a product, and of the product itself along with any by-products. If the totals do not tally, there must be a leak in the system. This systematic process reveals the full extent of the waste produced. It also provides statistics for working out how to reduce waste to the minimum in the future, and frequently results in major economic savings.

The wider scope of clean production

Clean production differs from such concepts as 'clean technology' or 'low-waste technology' in that it calls for more than the use of increasingly sophisticated technology. In the past the answer to a polluting process was, for instance, to install more efficient (and more expensive) filters. But this still left the problem of disposing of the contaminated pollution filter. A solution in accordance with clean production would find a raw material that did not create the problem in the first place. Clean production also demands that products are durable, reusable, and easily dismantled for reconditioning or to allow raw material to be recovered and used again; and it requires consideration of the eventual fate of a product to be given even at the design stage.

Clean production also focuses on the wider changes that are needed, such as ensuring that the right kind of educational and research facilities are available to industry. It even requires governments to place the central emphasis on preventing pollution, not on regulating waste disposal. In other words, governments as well as companies have to follow the waste minimization hierarchy.

GAS AND OIL
THE TRUE COST OF RICHES

THE DISCOVERY OF gas and oil under the North Sea in the 1960s and 1970s radically affected the economies of the countries that surround it. The North Sea now supplies 30 percent of these nations' energy needs, so creating huge savings in expenditure on imported oil. Industry has invested some $100 billion in exploiting the offshore oil and gas fields; governments have reaped roughly the same amount in taxes. More than 50,000 jobs depend directly on North Sea oil and gas.

Not surprisingly, the arrival of a vast new industry in the centre of the North Sea has not been without environmental costs. Every aspect of gas and oil production, from initial exploration to the fate of the rigs and pipelines once a field is exhausted, has consequences for the wildlife of the North Sea.

Crude statistics

Gas and oil are, generally speaking, found in different parts of the North Sea. The largest gas fields were created from coal deposited about 300 million years ago in the southern North Sea. They lie mainly under United Kingdom and Dutch waters, which are usually no deeper than 50 metres. The seams are about 6 kilometres below the seabed; the heat at that depth has forced gas out of the coal and it has then been trapped under layers of rock salt. Most of the oil has been distilled from marine deposits laid down roughly 150 million years ago, and lies under United Kingdom and Norwegian waters in the northern North Sea, which are deeper and rougher than those in the south.

Oil and gas are extracted from wells drilled from production platforms at sea and brought to land through more than 9,000 kilometres of pipeline. In the late 1980s the United Kingdom had a total of 92 platforms, the Netherlands 36, while 14 lay in Norwegian waters. Denmark had six platforms and West Germany two. The late 1980s were the time of peak of production, with around 130 million tonnes of oil and 73 billion cubic metres of gas being removed each year. If no

Exploration
In the shallow waters of the North Sea, 'jack-up' rigs (above) are used to explore the extent of the oil or gas reserves: the legs are lowered to the seabed from the floating platform, and the rig is jacked up into position. Exploration does cause some pollution — for example, oil-based muds that lubricate the drill bit contaminate the sea — but the main problems come when production starts.

Industry in mid sea
The Statfjord oil field in full production in the northern North Sea. Production platforms are huge structures, firmly grounded in water more than 100 metres deep. One platform will drill many wells, its well shafts radiating out in different directions to cover the whole area between it and the next platform in the field. All the oil extracted is piped first to one platform, and then to a terminal on shore. In some cases, the oil is pumped to a floating storage tanker and then shipped to shore.

major new fields are discovered, activity in the North Sea is expected to start falling off significantly after the year 2000, although production from smaller fields will probably continue for at least a century.

HOW THE INDUSTRY OPERATES

Oil and gas production falls into several stages: assessing the likely effects of exploiting a particular field, exploration, developing the production wells, extraction, transportation and refining, and final decommissioning of offshore installations. Each of these has implications for the environment.

Platform afloat
A production platform being towed to the Statfjord oil field. Drilling for oil in the North Sea started in 1972. Since then, both Norway and the United Kingdom have become self-sufficient in oil, but the cost has been a huge increase in pollution.

First considerations

Before even a drop of oil or a whiff of gas has been detected, decisions have to be taken about whether development is appropriate in a particular area. Although oil and gas companies have paid much lip service to environmental protection, in practice production has had priority. Even in the Wadden Sea, whose coastal wildlife is the area's richest and most vulnerable, gas is extracted close to the island of Ameland, in the heart of the Dutch Wadden Sea, while in the German area the one oil platform is still operating despite the risk of pollution.

Taking soundings

Extensive surveys are required to establish the extent and location of oil and gas reserves. From the pattern of echoes from explosions in the water, seismic surveyors map the structure of the underlying rock. The explosions are extremely loud, as they have to be powerful enough to penetrate deep into the rocks and return to the ship. In one piece of research, no whales, dolphins or porpoises – which are known to be sensitive to noise – could be found within 40 kilometres of such activities, even though they were normally abundant in the area.

The drills go down

If the geology of an area appears promising, a mobile exploration rig drills for oil or gas. If it is present in commercial quantities, a production platform then sinks wells down to the reserves. All drills use drilling 'muds' – a carefully formulated mixture of minerals and chemicals in a liquid base – which are pumped down the well shaft to lubricate the drill bit, carry the rock cuttings back up to the surface, and create a seal inside the drill shaft to prevent the oil or gas escaping. 'Deviated drilling' techniques make it possible for the well shaft to snake through the rocks, so that a single platform can drill up to 50 different wells. However, deviated drilling often uses oil-based muds, containing 10 percent or more of oil, to ease the passage of the drill and shaft. The mixture of mud and cuttings receives limited purification before being dumped overboard. This has been a huge new source of pollution in the North Sea. In 1988, 76 percent of the total oil pollution documented for the offshore industry, 22,555 tonnes, actually came from these oil-based muds, rather than from the North Sea oil and gas themselves.

The mud kills the fauna on the seabed under the rig, partly by smothering it and partly from the toxic effects of the oil. Within a radius of 500 metres from the rig, the normal community of animal life on the sea floor is replaced by a few specialized species that can cope with the polluted conditions. Similar changes continue up to 2,500 metres from the platform – an area of 20 square kilometres – and have been observed up to 5 kilometres away. Such effects can be seen even five years after dumping has ceased.

The original diesel-based muds rapidly killed marine life, and had been replaced by 'low toxicity' oil-based muds by 1985. But only a few species were tested with these alternatives, under a limited range of laboratory conditions. In actual use the 'low toxicity' muds have proved as toxic as the substances they replaced.

Emptying the mud tanks
Although the special 'muds' used in drilling could sometimes be reused, about 1,000 tonnes of mud will be dumped during the drilling of a single well, and just one platform may dump 4,000 tonnes of mud in a year.

Experiences such as these led the Norwegian government to demand that the offshore companies reduce dumping oil-based muds in Norwegian waters unless they can demonstrate conclusively that they are not responsible for the damage under the rigs. Others have been slower to take action.

At the well head

A working well brings a mixture of hydrocarbons and water to the well head. These are separated on the platform before the gas or oil is transported to shore and the remaining, so-called 'produced' water is discharged into the sea, although it still contains a considerable proportion of oil. For the North Sea as a whole, the amounts of oil discharged via 'produced' water increased by 63 percent between 1985 and 1988, to 4,096 tonnes.

'Produced' water is treated to reduce its oil content to 50 thousandths of a gram per litre or lower before being discharged. Once in the sea, this oil is diluted to around 1.6 millionths of a gram of oil per litre of sea water at a distance of 1 kilometre down-current of the discharge point. This itself adds to a general background level in North Sea water from all types of human activity of between 0.5 and 3 millionths of a gram per litre, and so raises the general concentration of oil to as high as 4.6 millionths of a gram per litre. These are apparently microscopic levels, but they are deeply worrying.

At levels of between only 5 and 20 millionths of a gram per litre, the zooplankton eat less and produce fewer eggs, many of which hatch into deformed young – and this can have adverse effects on the whole food web. The 'normal' background level and the point at which chronic long-term effects begin are far too close for comfort. The degree of toxicity may be related to volatile substances in oil known as polycyclic aromatic hydrocarbons (PAH). These are most abundant in newly discharged oil. Scientists are only just becoming aware of PAHs and, at present, PAH levels are not monitored.

Oil production in particular is an intrinsically messy business, and spills and leaks of oil happen almost continuously once a well is in operation. Oil spillage was thought to amount to some 500 tonnes per annum. But in 1987, a Dutch government aerial survey of the North Sea indicated that spills of oil from platforms could be as high as 35,000 tonnes every year. This would turn the offshore industry into one of the largest sources of oil pollution in the North Sea.

The subtleties of oil production, along with the need to maintain the platform and protect it from corrosion, mean that a huge range of anti-fouling chemicals, corrosion inhibitors, defoamers, detergents, dispersants, emulsifiers, gelling products, lubricants, oxygen scavengers, surfactants, viscosifiers, and weighting agents diffuse into the sea. But little is known about the amounts of these substances in the North Sea, and regulatory control is lax.

Floating giant
The Claymore A semi-submersible platform. This kind of rig is held in position by anchors on the seabed. A few semi-submersibles are used for production, but the deep rough waters of the North Sea mean that fixed platforms are the rule.

Fish on the line
An oil pipeline running across the seabed provides an angler fish with a perch. Fish living near production platforms absorb oil from the water, which their bodily defences then have to break down.

Life in the balance
Starfish, mussels and hydroids are growing in such profusion on the legs of this production platform (right) that it is impossible to see the underlying frame. The creatures settle on the rig as they would on a rocky seabed but, because they present a bigger surface to strong waves or currents, they put greater pressure on the rig and may topple it in a storm. For this reason, the production company may remove them with lethal anti-fouling chemicals.

A burning question
The oil that is brought to the well head is naturally mixed with gas, which is often burned off – releasing damaging hydrocarbons into the atmosphere (right). This is done because, in small fields at least, the profits to be made from collecting the gas are considered too small. One alternative to flaring is to reinject the gas into the well, leaving it there for possible use in the future.

A terminal effect
Most oil is piped to on-shore terminals, such as that at Sullom Voe in Shetland, for initial processing. This results in oil being discharged into the surrounding water, where it is absorbed by marine organisms, slowing their rate of growth.

Imperfect process
Oil refining (below) contaminates both air and water with volatile hydrocarbons – although scant attention has been paid to the quantities released into the atmosphere. The amounts discharged into the sea have been halved since the 1970s – but this is outweighed by the increased dumping of oil-based muds and 'produced' water from rigs.

Transportation and refining

Most oil and gas is transported by pipe to onshore terminal facilities. The largest is the Sullom Voe terminal in Shetland, which can handle a third of a million cubic metres of oil per day. Here, oil and water are separated further, and the oil is loaded onto tankers, to be taken to refineries elsewhere. Around 1,000 tonnes of oil gets into the sea from the terminals every year. Oil discharges from refineries have roughly halved during the 1980s, to around 4,000 tonnes a year.

But this level of fresh oil is still very high, given that most pours into confined estuaries and coastal waters. Refineries, along with all other sources of oil pollution, have raised the concentration of oil to between 5 and 25 millionths of a gram per litre in the coastal waters of the North Sea and as high as 60 millionths of a gram per litre in the inner estuaries. It would not be surprising if this oil had contributed, along with higher nutrient levels, to the changes in the plankton of coastal waters over the past decades. Certainly grazing copepods – the tiny animals that have a key role in the plankton food web – declined in the North Sea between the late 1940s and the 1970s, before apparently stabilizing in the 1980s. But the information simply does not exist to establish conclusively whether this is a natural trend or is due to human interference.

When the wells run dry

Some fields, platforms and pipelines are now approaching the end of their working lives. If abandoned they disintegrate, littering the seabed with wreckage. Governments have ordered that platforms will be entirely removed from the shallower southern North Sea, although it is not clear what will happen to the undersea piping, which is the greatest hazard to fishing. While structures in deeper waters must be removed to a depth of 75 metres, this still means that in very deep water the remains of rigs may jut 125 metres from the sea floor. The offshore industry has traditionally fought against the complete removal of rigs and pipes. The companies hint that the stumps of abandoned rigs in deep water would boost the fishing industry by providing an extra variety of habitats that would attract fish. But the number of fish would be small and their capture difficult. The industry's prime concern has appeared to be to save costs at the end of the production cycle, even though these are small in relation to the whole.

Wadden Sea conflict
Intense controversy has resulted in Germany from the building of this oil production platform in the Wadden Sea National Park. Although the rig has been surrounded with a steel wall, fishermen and wildlife agencies are still seriously concerned about the pollution it may cause in the event of an accident.

MINING THE SEA

The European construction industry uses gravel and sand in enormous quantities every year – and the North Sea has both, available for no more than the cost of digging them out and taking them ashore. Metals, too, are commercially 'mined' from the North Sea. All these activities may affect the fate of the area's wildlife.

The quantity of gravel being taken from the North Sea has increased dramatically since the 1960s: by 1988, 25 million tonnes were being taken annually. At this rate, reserves will be exhausted in about 40 years' time from the waters of the United Kingdom – the major extractor – and will last only 20 years, on average, in other areas. The reserves of sand are greater.

The species most at risk from this industry are the herring, which spawns on gravel, and the sandeel, which burrows in the sand to avoid its predators. The extractors should leave alone the sites that these fish are known to use – and should leave other areas untouched, too, to allow for changes in breeding rates and sites. Whenever sand or gravel is extracted, it should never be totally removed, and the surface should be kept flat. As reserves run low, there will almost certainly be increasing commercial pressure to dig deeper and remove gravel and sand with a far higher proportion of mud than is taken at present. This will create patches of turbid water, which may temporarily inhibit the growth of phytoplankton, which needs light to thrive. Worse still, heavy metals and other pollutants that have settled in the sediments over many decades will also be thrown up into the water once more.

The other main product of the seabed, currently exploited in France and Germany, is agricultural lime – which comes from sea shells and the remains of certain seaweeds. There are also potential sources of metal in North Sea sediments, including tin, chromium, zircon, thorium, yttrium, europium, and even gold. More remarkable is the 'mining' of the light metal magnesium from sea water. Norway, using its hydroelectric resources, produces some 60,000 tonnes annually at a plant near Oslo. Unfortunately the process creates 500 grams a year of highly toxic dioxins. Norway plans to reduce the emissions to 10 percent of 1989 levels by 1991 and to no more than 1–2 grams per year by 1994.

The oldest extraction industry of all in the area produces sea salt. Unlike other forms of mineral extraction, this is harmless to the environment. Indeed, far from creating pollution, its very survival depends upon the purity of the product – and of the sea itself.

Offshore disasters
A fire on the Piper Alpha platform in 1988, in which 167 people lost their lives, was fed by oil and gas piped from other production platforms, which could not rapidly close down their supply systems. Some 1,500 tonnes of oil were released into the sea, along with 4 tonnes of transformer oils containing PCBs – both potentially deadly to marine life. The Ekofisk Bravo blow-out of 1977 (inset) did not result in loss of life, but did release up to 22,000 tonnes of oil in seven and a half days. This spread through the water and the atmosphere before it reached the shore.

CATASTROPHES

Oil and gas rigs and platforms in the North Sea are built to withstand an enormous battering by wind and water. But no structure or system can be absolutely proof against natural disaster or human error. The most feared accident occurs when the head of a well fails to contain the enormous pressure under which oil and gas are forced to the surface. There has been one such blow-out, on the Norwegian Bravo platform in April 1977, which spilled between 15,000 and 22,000 tonnes of oil into the sea. Other major incidents resulted in the release of 3,000 tonnes of oil from a pipeline damaged by a ship's anchor in the United Kingdom Claymore field in 1986, and a further 1,500 tonnes spilled in the Piper Alpha disaster in the United Kingdom sector in 1988.

Oil slicks devastate rocky shores and are even worse for soft sediments and salt marshes. There, the oil may be absorbed into the sediment, creating a long-term source of problems. Even small discharges can cause many deaths. In 1987 over 10,000 birds were killed in the Dutch Wadden Sea by a spill of probably less than 1 tonne. A major accident in the northern oil fields at the wrong time could be just as disastrous. The risks are extreme in the north from the end of June until September when most young guillemots and razorbills, along with the flightless moulting adults, live in the same area as the platforms.

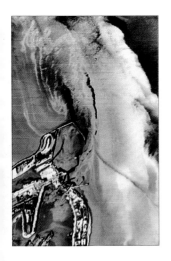

An estuarine slick
Computer imaging techniques expose an oil slick at a river mouth. Creatures living in the mud, providing vital food for wading birds, can remain impoverished for years because of the presence of persistent polyaromatic hydrocarbons in oil discharged from refineries.

Guillemots at sea
In the summer months, a large part of the auk population in the North Sea moves to the waters of the northern oil fields in search of food. A major oil spill at this time could have a devastating effect.

Oil is not the only potentially lethal substance that may be spilled in an accident. During the horrific Piper Alpha disaster over four tonnes of lethal PCB fluid from electrical transformers were lost. It was only this that alerted the regulatory authorities to the presence of PCBs in the middle of the North Sea – which says little for the foresight of the regulators. After April 1989, however, all PCBs were reported to have been removed from offshore installations.

THE FUTURE

North Sea oil and gas have brought major economic benefits to the people who live around its shores. But pollution has occurred that could easily have been avoided, for regulation has been weak and the industry has largely escaped the attention of the media.

There is no excuse for continuing to dump oil-based drilling muds at sea. Oil-based muds make the job only slightly faster than do water-based muds. Where, and if, the use of oil-based muds is unavoidable, the waste should be brought on shore. Experience in the Norwegian sector indicates that using water-based muds increases the total cost of drilling by only some 0.5 percent; bringing the mud ashore increases costs by only 0.2 percent.

The other major problem, of the oil discharged in 'produced' water and from terminals and refineries, is critical. Concentrations of oil in water have crept up to levels at which major effects might be expected to occur. It is essential that the oil discharged by the industry is reduced to an absolute minimum, and as part of a wider programme to reduce drastically all oil contamination of the North Sea. Again, experience in the Norwegian sector indicates that the cost of double- or triple-stage cleaning of 'produced' water, which would help to solve the problem, adds around 0.1 percent to the cost of each platform. And better solutions are unlikely to bankrupt the industry.

When oil and gas production ends, the companies should remove platforms, pipelines and all associated paraphernalia. There is no reason why the offshore operators should not leave the North Sea in as good a state as they found it.

The price of oil
Oiled seabirds die from exposure because their feathers are no longer waterproof, and from the toxic effects of the oil they ingest when they try to clean themselves.

SHIPPING
FRAUGHT WITH DANGER

NORTHERN EUROPE is the most concentrated area of industrial activity in the world, and its commercial facilities are clustered around three sides of the North Sea. There is a massive distribution by ship of raw materials and finished goods among the North Sea states and further afield. Each year, almost half a million voyages take place in the southern North Sea alone.

Accidents and careless operating practices have given shipping a bad record of pollution, although there is no intrinsic reason why this cannot be improved. And a unique environmental conflict is created by one distinct aspect of transportation, below, on and above the surface of the North Sea – that of the military.

ACCIDENTS AT SEA

By 1967, traffic in the Dover Strait was so great that a scheme was introduced to separate ships travelling east from those travelling west. But the scheme was only voluntary and, inevitably, disaster struck. In 1971, the *Texaco Caribbean* sank after colliding with the *Paracas*, a Peruvian ship going against the flow of traffic; the German ship *Brandenburg* collided with the wreckage, and was itself hit by the Greek-registered *Nikki*. Many people lost their lives. After the tragedy, radar surveillance was introduced into the area; but it is still perfectly lawful to travel in the wrong direction down the busiest shipping lane in the world.

The separation scheme now extends into Dutch waters – 60 percent of all North Sea shipping passes through here. A complex set of routes weaves its way through gas and oil fields, fishing grounds, ocean incineration areas, sand and gravel extraction sites, and military exercise zones. But the segregation is not total and, in some areas, oil and gas platforms lie in the middle of the shipping lanes: in the late 1980s, the *Vinca Gorton* collided with a major gas pipeline, causing damage that cost more than £10 million.

Who pays the bill?
When the *Mont Louis* sank off Ostend in Belgium in 1984, she was carrying a cargo of radioactive, chemically explosive and toxic uranium hexafluoride from Le Havre to the Soviet port of Riga, where it was to be processed for use in French nuclear reactors. The salvage operation was left to the Belgians, who recovered all 30 containers. Except in the case of oil pollution, those responsible for hazardous cargoes have so far refused to accept that they should pay for salvage operations after an accident, or that they should compensate for any damage.

New ships, old problems
Cargo ships have changed considerably over the past 30 years. Instead of the cargo being stowed away piece by piece at the port, goods now arrive ready packed in containers, which are loaded onto container ships (left) or driven onto roll-on roll-off cargo or passenger ferries. All in all, the use of these new ships, fitted with sophisticated safety and navigation equipment, has – surprisingly – not been matched by a corresponding reduction in the loss of ships or cargoes.

Throughout the southern North Sea, huge numbers of ferries add to the traffic problems, because their routes cut right across the shipping lanes. The French-flagged *Mont Louis*, carrying a cargo of radioactive uranium hexafluoride, sank off the Belgian coast in 1984, and the Dutch-flagged chemical tanker *Anna Broere* went down with a cargo of vinyl cyanide (used to produce synthetic fabrics) off the eastern coast of England in 1988 – both having collided with ferries.

Caught in a storm

A combination of bad weather and the failure of the steering gear caused the tanker *Amoco Cadiz*, flying the Liberian flag, to run aground on the coast of Brittany in 1978: the ship broke up and its entire cargo of 223,000 tonnes of crude oil poured into the sea. Huge numbers of shellfish, birds and other wildlife on the beaches of northern Brittany were contaminated. Oil penetrated estuaries and inlets, too, sinking into the sediments there and creating a hazard to wildlife that persisted long after the accident.

Agricultural chemicals are among the most dangerous cargoes lost at sea, as these are often acutely toxic to marine life. In 1984 the *Dana Optima*, travelling from Newcastle to Esbjerg under the Danish flag, lost its deck cargo in hurricane-force gales. The load included 16 tonnes of dinitrophenol, a herbicide banned in the United States because at concentrations of only 10 thousandths of a gram per kilogram of body weight it causes reproductive failure in mammals. Despite determined efforts to recover the load, 1.8 tonnes of dinitrophenol still lie in the North Sea, and the containers are slowly corroding.

The *Perintis*, registered in Panama, was carrying 5 tonnes of the pesticide lindane from Antwerp to Jakarta when it went down off the Channel Isles in 1989. The lindane was never found, despite a search by a French submarine. The Dutch government has calculated a tentative 'safe' level for lindane in the North Sea of only 2 billionths of a gram per litre. If the 5 tonnes from the *Perintis* were spread throughout the top 100 metres of the North Sea and English Channel, the pesticide would have a concentration of 0.15 billionths of a gram per litre – just

The price of oil
The supertanker *Amoco Cadiz* was wrecked in March 1978, contaminating huge stretches of the Brittany coast with crude oil. The cost of cleaning up after the disaster reached US$ 2 billion, and the effects of the oil persisted for years – especially in the sand and mud of the estuaries.

False economy
The *Amoco Cadiz* was lost when the rudder failed and tugs could not stop her from drifting onto rocks. The failure occurred because routine oil changes had been ignored. The few gallons of oil saved led to 223,000 tonnes being spilled into the sea.

Hazardous cargo
A consignment of dangerous chemicals is examined for leaks. The cargo of epichlorhydrine was damaged in rough weather during the voyage from Russia to Rotterdam. Huge quantities of hazardous substances are regularly transported by sea, and accidents happen far too often.

13 times less than the 'safe' limit. But if the lindane does leak, it will be dispersed through a far smaller area, passing up the English Channel and along the shallow waters of the coast of northwest Europe. Some argue that the chemical will be released into the sea only very slowly, and that it will break down to form harmless substances. Others are concerned that either bacteria in the sediments or ultra-violet light from the sun may convert it into even more toxic substances that accumulate in living organisms with lethal effect.

DAY-TO-DAY PROBLEMS

Even in everyday operations, a ship's engine generates large amounts of oily waste. Ships also fill empty fuel and cargo tanks with sea water to act as ballast, or to wash out residues of chemicals. All too often these wastes are discharged into the sea. Illegal oil discharges are monitored from the air, but only sporadically. So far, the North Sea states have been unable to agree on how much oil is discharged; estimates range from 1,100 to 60,000 tonnes per annum – which means that the pollution could be on the same scale as that from the entire offshore oil and gas industry. No measurements have been made of chemical discharges.

Most ports in the North Sea now provide facilities for the disposal of waste oil and chemicals, although many are inadequate – or are not used. For instance, in 1988, three Belgian ports – Antwerp, Ghent and Zeebrugge – handled 26,095 tonnes of waste oil, whereas they had expected 300,800 tonnes; there was a similar discrepancy in the amount of chemical waste handled.

The dumping of rubbish

Much of the litter that turns up on beaches around the North Sea comes from shipping. All of it is unsightly, but items such as discarded fishing net and line and the plastic yokes from multiple packs of canned drink entangle seabirds and kill them. Birds will also swallow plastic pellets, which accumulate in their crops, limiting the amount they can eat.

Some ships, particularly ferries, now collect their rubbish and dispose of it when they reach port. But it has taken the North Sea states years to agree that the littering of the sea is unacceptable, and it will not become illegal until 1991. The difficulty in reaching agreement on such a straightforward issue is a graphic illustration of the inertia that affects the regulation of shipping.

ROUTES TO A CURE

Danger on the shore
It is essential that potentially dangerous cargoes are carefully labelled. Otherwise, should they get washed overboard, those who have to deal with them when they come ashore face enormous problems in identifying them. But it is just as important to improve operating practices so that the amount of cargo lost at sea is drastically reduced.

Given the technical advances in navigational aids over the past 30 years, it is shocking to find that there has not been a dramatic decline in accidents at sea. In fact, between the early 1950s and the late 1970s the percentage of ships lost worldwide declined only slightly from 0.71 percent to 0.56 percent, and the tonnage lost – the combined weight of ship and cargo – actually increased from 0.31 to 0.49 percent. This has happened because the number of crew have been reduced, and levels of training are low, and because navigational and safety equipment is either not installed, or is incorrectly used or broken.

Some states (outside the North Sea area) have very poor operating standards, but improvements can only be insisted upon with the agreement of all nations. The North Sea states have argued that this system prevents them improving the practices of ships that operate in their waters, although the specific issues of pollution from waste discharges and navigational standards in their waters are now coming increasingly under their control.

But just a handful of measures would make the North Sea a much safer place. Day-to-day pollution could be reduced by providing better facilities in ports, along with the imposition of severe penalties for discharging waste at sea. In the longer term, the design of ships could be improved, to reduce the amount of

The healthy sea air?
The atmospheric pollution produced by shipping is greater than that produced by some North Sea states. The sulphur content of the fuel oil used by long-haul steam turbine ships is high, around 3.5 percent, all of which is discharged into the atmosphere when the fuel is burned. As a result, around 100,000 tonnes of sulphur dioxide are released over the North Sea every year. Emissions of nitrous oxide from turbine ships and short-haul diesel-powered vessels are also significant, in the region of 20,000 tonnes per year. The pollution has the biggest impact in enclosed areas such as the Norwegian fjords or in crowded stretches of the sea such as the Strait of Dover.

waste oil they produce. There should also be a tighter, mandatory, system of traffic control – perhaps an adaptation of that used for aircraft – which would reduce the number of shipping accidents dramatically. It would also end the insanity that allows ships to progress against the flow of traffic.

Lost cargoes could be made less of a problem to the environment. Only waterproof containers for hazardous chemicals should be used and they should be resistant to corrosion; automatic flotation devices could be built into them, as could small radio beacons to make them easier to find once they have been lost.

Where damage is caused, full reparation should be made by those responsible, who should have the necessary insurance to cover the cost. Although shipping companies do sometimes pay compensation for polluting the sea with oil, industry has so far strongly resisted the idea that any compensation should be paid for the pollution caused by pesticides and other harmful substances – or that the companies responsible for the loss of chemicals at sea should pay the price of recovering them.

The environmental problems created by the shipping industry could be solved readily, with relatively few technical difficulties. But for this to happen the people who work in shipping must change their attitude to the sea, and stop thinking of it as a bottomless pit into which they can discharge all manner of substances without ill effects. That may take a little longer to achieve.

Seal casualty
The wildlife of the North Sea can suffer terribly at the hands of the shipping industry. The large wound in this harbour seal's side is believed to have been caused by a ship's propeller.

WILDLIFE AND THE ARMED FORCES

A large amount of the hazardous material washed up on the shores of the North Sea has a military origin. The debris – the legacy of two world wars – ranges from individual mines and bombs to the chemical weapons dumped in the Skagerrak and Kattegat after the 1939–45 conflict, and the sunken munitions ships scattered throughout the sea – from which all keep a wary distance.

However, a quarter of all hazardous items washed on to the beaches of the United Kingdom comes from military activity taking place today, and there is no reason to believe that the situation is substantially different in other countries.

The North Sea is intensively used by the military. The Soviet navy enters it both from the Baltic and from the north, partly to monitor NATO activity. In the northern North Sea there are regular NATO naval exercises involving many ships that are known or can be assumed to be carrying nuclear weapons. NATO also uses the North Sea to monitor Soviet submarines passing through the northeast Atlantic, involving both anti-submarine aircraft operating from bases in Scotland and Norway, and anti-submarine ships in the sea's northern waters. In addition there are constant day-to-day activities by the navies, air forces and armies. Accidents occur frequently, ranging from the loss of munitions to aircraft crashes, and the risk of catastrophe, such as a collision between a submarine and an oil platform, is always present.

Of particular concern is the operation of nuclear submarines, which do not have a good safety record. Since they came into use, 6 submarines have sunk, resulting in the loss of 7 nuclear reactors and at least 36 nuclear warheads. None has so far been lost in the North Sea, but the threat is clear. United Kingdom nuclear submarines make frequent journeys to the repair and maintenance yards at Rosyth, in the Forth estuary; both NATO and Soviet nuclear-powered hunter-killer submarines lurk in the deep waters of the Norwegian Trench; and French nuclear submarines of all types use the waters of the Hurd Deep for access to the naval docks at Cherbourg.

Numerous areas are set aside for military exercises throughout the North Sea. Most controversial are those sites in the national parks of the Wadden Sea. For example, in the Schleswig–Holstein area there is a weapons-testing range for both the German army and commercial manufacturers, which surrounds the island of Trischen, a sanctuary for birds. In August each year, 100,000 shelduck – the majority of the North Sea population – come here to moult, during which time they are unable to fly and are therefore very vulnerable. The shelduck (and other birds) panic – not only during the firing exercises themselves, but also when helicopters fly over to check that nobody is in the target area before firing begins, and to recover the projectiles afterwards.

Nuclear visitation
A United States nuclear submarine on a courtesy visit to Rotterdam. Civic authorities allow military nuclear power into densely populated urban areas, where a civil nuclear power plant would be unthinkable.

Seals that use the sand banks in the area are also disturbed. The impact that military exercises have on seals is clear from what happened on the island of San Miguel, off the Californian coast, when the US military abandoned an exercise zone. Once the military explosions stopped, seals began to colonize the area, and there are now more than 100,000 of them, of five different species. There is no reason to believe the situation would be any different in the North Sea.

Other military areas in the Wadden Sea include an aircraft firing range on the island of Sylt, a shooting range at Rømø in the Danish sector, and an area of intensive army, navy and air force activity in the eastern Dutch Wadden Sea around Den Helder and the island of Vlieland. There are also exercise areas on the UK coast of the North Sea and in the English Channel.

The military sometimes argue that their activities actually benefit wildlife because they restrict human access. But there are other, far more rational and more cost-effective, ways of protecting animals, without the use of explosives! If governments were to place the same emphasis on environmental security as they have on military security in the past, then we would see an end to the activities that risk damage to wildlife and people alike.

Military zone or park?
Wreckage and ammunition is recovered (left) after a firing exercise around the island of Trischen, in the German Wadden Sea. The military use of Trischen's mud flats is in direct conflict with the area's status as a national park.

Moulting grounds
An aerial view of part of a vast flock of 100,000 moulting shelduck (*Tadorna tadorna*) on mud flats in the German Wadden Sea. In August the shelduck congregate here from both the west and east coasts of the North Sea. During the moult, the birds are unable to fly and become distressed by military activity that takes place in their vicinity.

PART 3

In the past 10 years, Greenpeace's efforts in the North Sea have all been directed towards eliminating our persistent damage to the sea's rich yet precariously balanced ecology. Greenpeace's first move in tackling a specific issue is to establish that a particular activity is indeed harming the environment. Armed with this information, they will then approach governments or national or international agencies to put the case for change. If that fails, Greenpeace will go direct to those causing the problem. Only after these three moves have been exhausted will the activists plan a spectacular exposure of an environmental hazard or abuse – in the hope that publicity will create fresh opportunities for dialogue with those ultimately responsible.

Direct action has been part of the Greenpeace story from the very beginning. It derives from the Quaker principle of 'bearing witness' – peacefully registering one's protest simply by being present at a deplorable activity. Greenpeace activists have certainly done more than that – by chaining themselves to ships dumping noxious waste into the North Sea, by blocking pipes and scaling chimneys. But they have also done a great deal more, by tackling politicians, bureaucrats and the abusers themselves with the glaring evidence of their failure to look beyond their own immediate interests.

The battle for the sea
Crew members of the incineration vessel *Vulcanus* turn high-pressure hoses on Greenpeace activists trying to persuade them to stop burning toxic waste from Spain in the North Sea. The decision by North Sea ministers to prohibit ocean incineration from the end of 1991 marked a major victory for Greenpeace's eight-year-long campaign.

GREENPEACE
A DECADE OF CAMPAIGNS

AT THE BEGINNING of the 1990s, Greenpeace's efforts to bring attention to the plight of the North Sea are remarkable for their organization, the wealth of scientific research that backs them up, and their success in getting their message home. It is difficult to believe that at the start of the very first North Sea campaign in 1980 Greenpeace had no clear idea of what specific issues to concentrate on, other than the intuitive certainty that the North Sea was being damaged by a flood of toxic waste that European industry was dumping into the sea. Greenpeace activists candidly admit that the North Sea campaign was born more of inspiration than calculation or research.

In 1980, there were five Greenpeace offices in Europe, all newly opened – in the UK, the Netherlands, West Germany, France and Denmark. The problem of toxic waste was an issue that all five countries had in common. Extraordinary as it seems now, there was at that time little or no public awareness that the North Sea had become Europe's industrial dustbin. The question was: which particular form of waste dumping was to be Greenpeace's target?

Before that question had been answered, the Dutch Greenpeace office learned in the spring of 1980 that liquid waste containing lead, cadmium and mercury was regularly brought in barges down the Rhine from dye factories at Leverkusen belonging to the giant German chemical company Bayer, to be discharged into the sea from two dump ships. Somehow, Bayer learned when Greenpeace planned to act against the ships, in Rotterdam harbour, but made no attempt to divert them. Greenpeace activists proceeded to immobilize the two dump ships and a barge laden with waste by chaining inflatable dinghies to the rudders and discharge

Official secret – exposed
Greenpeace inform citizens of Hamburg in 1989 that the visiting HMS *Ark Royal* has more than aircraft on board. Greenpeace's campaign against nuclear contamination of the North Sea began in 1980. It has often focused on the risk of accident from the nuclear-armed ships and nuclear-powered submarines that regularly use the sea, and successfully brought to an end the dumping of nuclear waste in the Atlantic. Greenpeace have also drawn attention to the fact that the North Sea is already polluted by radioactive waste – brought there by ocean currents, from the nuclear fuel reprocessing plants at Sellafield and Dounreay in the UK and at Cap de la Hague in France.

Seeing and believing
A Greenpeace campaigner takes a sample from the pipeline disgorging toxic waste into Antwerp harbour from the nearby Monsanto pesticide factory. Behind every Greenpeace campaign and direct action in the North Sea lies a solid body of scientific research, which activists use both to alert the public to what is happening around their shores and to prove to governments and industry that their policies are ultimately destructive.

pipes. All three vessels remained paralysed for three days before Bayer took out an injunction against Greenpeace, and the organization – and the issue of toxic dumping – basked in spectacular publicity.

But then Greenpeace learned that the dumping was entirely legal – the Dutch government issued permits regularly for the operation – and that all the various chemicals being taken from Rotterdam harbour were first stored in tanks, creating a horrifying brew of poisons whose interactions could scarcely be guessed at.

Greenpeace's reaction was to ask by what loophole heavy metals that were prohibited from being dumped at sea by the Oslo Convention of 1972 were legally being discharged into the ocean. The answer was that these were regarded as 'trace' metals in the waste – for no one had defined what proportion of the waste constituted a trace. The activists determined to establish exactly what damage was being done to the life in the North Sea – and to look all the more urgently for a single issue on which to focus a campaign of protest.

A campaign is born

Then, Greenpeace supporters in Hamburg blockaded ships taking titanium dioxide waste from factories belonging to the American-owned company Kronos-Titan out of Bremerhaven. This action was the inspiration for the first sustained Greenpeace campaign in the North Sea. Of the five countries where Greenpeace were active, all except Denmark had large titanium dioxide plants. These were dumping 5.6 million tonnes of waste into the sea every year.

Little was then known of the precise effects that titanium dioxide waste had on life in the sea. The chemical had been used as a whitener in paint since the 1960s, when it replaced poisonous lead additives. But the process used to extract titanium dioxide from its ore was largely inefficient, and the waste it produced contained acids and heavy metals: when dumped, it stained the sea white.

Greenpeace adopted the campaign in December 1980, immediately issued the demand that dumping titanium dioxide waste should cease by the end of 1985, and began to collect evidence of its effects on life in the sea. Greenpeace researchers discovered that, in 1972, an Italian factory had been found pouring undiluted titanium dioxide waste into the Mediterranean off Corsica; a massive public protest had led to the French courts ruling that the dumping was illegal. Since then, two things had happened. Numerous scientists had identified grim effects on fish exposed to the waste and had aired their strong suspicion that the poisonous ingredient was the high proportion of ferrous sulphate in the waste. And in 1978 the European Commission had issued a directive, which had never been acted on, to end dumping. Greenpeace funded further research, maintained the campaign of direct actions, pressed for early legislation by lobbying the European Commission and the Oslo Convention, and tried to persuade the two major ocean dumpers, Bayer and Kronos-Titan, to see the error of their ways.

Greenpeace also began a legal campaign at that time. Working in a coalition with two Dutch environmental groups, Nature and Environment and the North Sea Working Group, the activists appealed to the Dutch Council of State against decisions to renew the dumping permits, backing their argument with scientific evidence. Those appeals failed, but by 1982 the Dutch were granting dumping permits valid for only a year and had announced that Kronos-Titan and another producer, Pigment Chemie of Belgium, would have to halve their dumping of titanium dioxide waste, while Germany had announced that ferrous sulphate dumping would be banned from 1984. The companies involved were also beginning to respond. Bayer in Germany, for example, brought in cleaner manufacturing techniques three or four years ahead of schedule and announced an end to their North Sea dumping in June 1981. In just two years, Greenpeace's campaign was getting positive results.

Fighting the white menace
Crew members of the *Falco*, which was dumping titanium dioxide waste from the Belgian NL Industries' factory in Ghent, attempt in 1983 to cut away a Greenpeace protester who has chained himself to the ship next to the discharge pipe. Greenpeace actions against titanium dioxide dumping began in 1980. The campaign, which also involved scientific research and lobbying governments, regulatory bodies and manufacturers such as Pigment Chemie, Bayer and Kronos-Titan (owned by NL Industries), was Greenpeace's first sustained effort on behalf of the North Sea.

A hard rain
Greenpeace helped to bring acid rain to public attention in the early 1980s. It is produced in northern Europe by pollution from burning fossil fuels, and enters the North Sea from the atmosphere. The most spectacular action, in April 1984, involved activists climbing eight smokestacks in eight countries and hanging a huge banner on each one. Each banner showed the legend STOP ACID RAIN and one vast letter. When photographs of the chimneys were placed side by side, the letters spelled out STOP STOP.

Antwerp blockade
Greenpeace blockade the dump ship *Wadsy Tanker* in Antwerp harbour in April 1985. In May, the *Sirius* was impounded after blocking the harbour. The *Sirius* finally escaped by canal into Holland, but only after the top third of the ship had been removed so that it could pass under the canals' low bridges.

Danger from gypsum
In June 1983, Greenpeace protesters try to prevent a barge releasing gypsum sludge waste, from a fertilizer factory, into the sea.

New issues, new actions

By 1983, after changes within Greenpeace, the organization had broadened the scope of activities and targets across a range of issues in the North Sea – from ocean incineration and the French dumping of gypsum sludge to the effects of PCBs and the British dumping of sewage. A number of *ad hoc* actions took place over the next two years against such targets as the three French companies Cofaz, Rhône-Poulenc and APC, who were dumping 2 million tonnes of gypsum sludge, containing cadmium, from their fertilizer factories on the Seine.

And the campaign against dumping titanium dioxide waste at sea continued to gain ground. At the end of 1984, Pigment Chemie and Kronos-Titan announced that they were building recycling plants and would cease dumping in 1988.

Bayer's Belgian plant in Antwerp, however, continued fouling the North Sea, claiming that the only alternative was to close the factory, and Belgium remained the only country in Europe to have no controls on dumping in the sea. Fearing that Belgium would become the conduit for waste illegal elsewhere, Greenpeace intensified their campaign in Belgium. In April 1985, ships running waste from Bayer plants and from NL Chemicals of Ghent were boarded or harassed; and in May, the Greenpeace ship *Sirius* blockaded Antwerp harbour, closing part of the port to all shipping for 48 hours. Five months later, the Belgian government announced that it would ban dump ships from its harbours by the end of 1987 – although, in the event, the ban was not actually enforced until 1989.

THE SECOND NORTH SEA CONFERENCE

It was not until 1986, when Greenpeace found themselves on a much sounder financial footing and had further rationalized their strategies, that a fully co-ordinated North Sea campaign began. By then, also, there were Greenpeace offices in all of the North Sea states except Norway. The campaign was to embrace the whole range of problems that the sea faced, and it was to become the model for all the regional campaigns that Greenpeace have launched since.

In pursuit of the British
Greenpeace dinghies pursue the *Yarrow* as it leaves Teesside in April 1987 to dump its cargo of industrial waste in mid ocean. The action was part of the campaign to expose the UK as 'the dirty man of Europe'. First Greenpeace delayed the *Yarrow*'s departure, then local people invaded the ship. Greenpeace activists tied a steel cable to the propeller and made it fast to the quay. The *Yarrow* left next morning, only to be harried by Greenpeace's *Sirius* for 12 hours, during which activists boarded the vessel. The *Yarrow* finally turned back – still fully laden.

The immediate concern in 1986, however, was to make a concentrated effort to influence the meeting, due to take place in London in 1987, of the environment ministers of all the North Sea states at the Second International Conference on the Protection of the North Sea – the 'North Sea conference'.

Greenpeace were especially cynical about the role of the United Kingdom at the conference, for the government gave every impression of being indifferent to the environment. The suspicion was growing that the UK had agreed to host the second North Sea conference with the intention of diluting its aims.

Whereas much of the rest of Europe had begun to recognize that the only sensible approach to pollution control was to attack the problem at source, the UK insisted that scientists could establish the level of pollution that the sea itself could bear. If such a level were exceeded, corrective measures would be taken.

The UK government also went so far as to cast doubt on any connection between industrial waste and the alleged collapse of the North Sea's ecology, and demanded that scientists first show an unambiguous relation between the two. But this policy would mean that remedial action would always come too late, after the damage was done – perhaps irreversibly and irredeemably.

Britain stands alone

Greenpeace now planned to expose the United Kingdom as the dirty man of Europe, with the intention of making the political pressure so intense that it would be forced to revise its position. At the sharp end, as always, of Greenpeace's campaign would be direct actions, and huge resources of energy and ingenuity were devoted to making them as effective as possible.

Greenpeace were also intensifying their own research, using the newly commissioned laboratory ship *Beluga*, and in March 1987 published a report that revealed that half of the dab and flounder caught in parts of the North Sea were diseased. Another Greenpeace report disclosed for the first time that many fish in the Thames, which was reputedly cleaner than it had been for a century, were also badly diseased.

The politicians of Europe were at that time looking at the draft of another alarming report on the North Sea, one that they could not easily deny. This was their own 'Quality Status Report', which attempted a systematic evaluation of the health of the North Sea, in preparation for the forthcoming conference. Despite its weaknesses and signs of political compromise, the report left no doubt that the waters of the North Sea were heavily polluted and that plants and animal life were either badly affected by the contamination or were at risk from it.

The first conference
Greenpeace hang out a reminder to the ministers attending the first North Sea conference at Bremen in 1984. Greenpeace judged that the forum would achieve little – and, indeed, although the conference agreed to take steps to maintain the quality of the sea, no deadlines were agreed, and no specific targets were set.

As some of the politicians later explained, the report also made it clear to them, once and for all, that science could never determine exactly how much pollution the North Sea could absorb. And so they decided on a new approach: the adoption of the 'precautionary principle', which stated that the level in the sea of any substance suspected of being harmful to the environment should be reduced, even in the absence of scientific proof. In other words, for the very first time, the environment was to be given the benefit of the doubt. A sign of the change in mood came when the international Rhine Commission agreed that the most harmful substances in the river should be reduced by 50 percent.

In the months leading up to the conference, Greenpeace were carrying on their relentless campaign of direct actions, accompanied by the *Beluga*'s analysis of all of the major estuaries feeding the North Sea, which had been carried out in the summer of 1986. In each country, Greenpeace argued that the estuaries were seriously under threat from pollution, and that it was the UK that was blocking progress to change. Other research supported their own – for instance, a University of Hamburg survey estimated that in the waters of Rotterdam harbour alone, there were 60,000 chemicals in more than a million combinations.

By now there was a strong swell of public opinion that something should be done to save the North Sea. About a month before the second conference, the Wadden Sea states met privately to agree a common position: to argue for the adoption of the precautionary principle, and to press for reductions in the quantities of some chemicals poured into the North Sea. Meanwhile, the Swedish government had concluded that specific measures had to be agreed at the conference if disaster – both environmental and political – were to be avoided. They had argued, as had the Rhine Commission, since July 1987 that they wanted the most dangerous substances, and nutrients, to be reduced by 50 percent. Other Nordic countries and the Environmental Commission of the EEC supported them.

A living curtain for the Rhine
Greenpeace activists form a screen of bodies across the Rhine in December 1986, by dangling from the Leverkusen Bridge. The action was designed to highlight the continued toxic pollution of the river by the giant chemical company Bayer.

Welcome to the conference
Greenpeace activists line up barrels of toxic waste from the North Sea states to greet the delegates at the opening of the second North Sea conference in London on 24 November 1987. Campaigners wearing protective clothing had surreptitiously taken the waste directly from factory pipelines discharging into the sea.

But even on the eve of the conference it was still not certain what the outcome would be. Many expected that it would achieve nothing, because the UK was proving so intransigent, and the Nordic nations told Greenpeace privately that they were prepared to make their dissatisfaction plain by walking out.

However, two things happened at the conference that changed the position. First, it became clear that neither the French (who saw the North Sea as a very peripheral issue) nor the Belgians (who were engaged in an internal political dispute over whether the national or regional governments had responsibility for the environment) were prepared to support the UK.

The second event was more unexpected. The opening of the conference in London on 14 November 1987 was marked by a surprising speech from the Prince of Wales. Prince Charles spelled out his message in a simple metaphor: 'While we wait for the doctor's diagnosis, the patient may easily die.' The Prince's words, and their coverage in the British press, had one crucial effect: they convinced the UK delegation that politically it was completely isolated.

The next day the UK government beat a strategic retreat and signed a document proclaiming the precautionary principle as the guiding policy for the North Sea, and agreeing to implement it by reducing nutrients and especially dangerous substances in the sea by 50 percent of their 1985 levels by 1995. A date for ending ocean incineration was also agreed. Greenpeace, which had first brought these issues to public attention, could justifiably claim that they had won a major political battle – although they had no doubt that the struggle to give the North Sea a chance to recover its health was far from over.

Where there's smoke...
Drastic mistakes were made when burning toxic waste at sea. In theory, the incineration ships' furnaces generated such high temperatures that the waste burned in a flameless fire, and residual ash represented less than 0.1 percent of the original cargo. The facts were less palatable. The white smoke the ships generated was vaporized hydrochloric acid still contaminated with HHCs. When the ships' furnaces developed faults, the smokestacks would belch giant flames and a shower of sparks, followed by a gigantic plume of black smoke – full of unburned particles of poisonous waste.

Contaminated by fire

During this period the campaign against the incineration of hazardous waste at sea came into its own. The campaign had been launched in 1982 when the discovery that the incineration vessel *Matthias II* was emitting deadly dioxins provoked Greenpeace to call for the ship to be withdrawn from the North Sea.

During the 1980s, incinerator vessels burned a cocktail of noxious, persistent chemicals, considered far too dangerous and corrosive to be incinerated on land. Waste from the manufacture and use of PVC, pesticides, pharmaceuticals and organochlorine solvents supplied two burn ships with an annual load of 100,000 tonnes to be incinerated in the North Sea.

Although ocean incineration had been labelled an 'interim' method of waste disposal, governments only began serious discussions about bringing it to an end when challenged by Greenpeace. In March 1987, two Greenpeace reports cast new doubts on the environmental acceptability of ocean incineration. One showed that the theory underlying the process of incineration was flawed, and called into question the information used to support incineration over the past 20 years. The other report provided evidence that emissions from an incinerator could harm organisms, including fish larvae, that feed near the very surface of the ocean. Additional evidence, released by West Germany, that hazardous chemicals had been found in the sediment and fish of the burn zone led to the decision at the second North Sea conference to stop ocean incineration by the end of 1994.

By early 1989, pressure on the North Sea states, and on the administering Oslo Commission and London Dumping Convention, to bring forward the date had reached an extraordinary pitch. The first country to give way was Belgium, which in October banned the import of waste for incineration at sea. As many countries shipped their waste through Antwerp, this proved a fatal blow. Then West Germany disclosed that waste awaiting incineration was contaminated with dioxins – substances so toxic and persistent that it was illegal to burn them at sea. Under intense public pressure, the government drastically reduced the amount of waste that could be burned and, soon afterwards, the owners of the burn ship, the *Vesta*, withdrew her from the North Sea – ending incineration of toxic waste from the German region.

By the end of 1989, only the UK, Spain, Belgium and Switzerland were left burning waste at sea. Ministers of those states, too, have now agreed to end the operation entirely by the end of 1991.

During the campaign, the polluters had challenged Greenpeace to suggest alternative or more efficient means of disposing of the waste. But Greenpeace had repeatedly pointed out that the only solution was to eliminate toxic waste altogether; more generally, the amount of all waste produced had to be reduced to a minimum. The organization became increasingly active in its promotion of 'clean production' – which, briefly expressed, involves developing manufacturing techniques that emphasize the avoidance of waste creation rather than waste disposal – and the practice is now a major part of their thinking.

Gathering fuel
In 1987, as the *Wedau* gathers its unwholesome cargo of toxic chemicals from factory after factory along the Rhine for burning at sea, Greenpeace advertise the waste ship's presence with huge banners. Over the four days of the action, Greenpeace set in motion a massive telephone campaign, to tell the people who lived or worked along the Rhine what was passing their doors so regularly, and pointing out what might happen if the *Wedau* should be involved in an accident. Perhaps more than anything else this action aroused public awareness, and ire, in West Germany, Holland and Belgium over the ocean incineration programme.

Suffering of the innocents

In the summer of 1988, about 50 percent of the North Sea's population of harbour seals – more than 18,000 animals – fell victim to a mysterious disease. Greenpeace's European offices called on emergency funds and immediately began a series of actions to publicize the seals' plight throughout the North Sea states, and they set up a conference of scientists to discuss the possible causes.

The scientists concluded that the immediate cause of the seal deaths was a virus; and, although they had no proof, they strongly suspected that the epidemic was linked to pollution. Greenpeace pointed out forcefully that if the countries of the North Sea had any faith at all in their own precautionary principle, they should stop contaminating the sea at once. If only one nation would put the precautionary principle into practice before the third North Sea conference, the political effect would be enormous.

The North Sea states did call an immediate halt to harbour seal culling and agreed that further reductions in pollution entering the sea would be tabled at the third North Sea conference in 1990. The seal deaths were also crucial in convincing the Nordic bloc that only the most drastic action would save the sea; and, as preparations for the 1990 conference began, the Nordic countries became increasingly sympathetic to Greenpeace's position.

TIGHTENING THE SCREWS

The third North Sea conference was scheduled to take place at The Hague in March 1990. While direct actions remained the priority, to maintain public awareness of the issues at stake, Greenpeace recognized that the political arguments at the conference would be won or lost as much on scientific grounds as through environmental activism, and planned accordingly.

Dead in Downing Street
A victim of the 1988 seal plague outside the British prime minister's London residence. Scientists at a conference organized by Greenpeace had suggested that without the pollution in the sea, the epidemic might not have been so severe. The UK's response was to maintain that the case against pollution remained unproved.

Sampling the water
A Greenpeace volunteer (left) risks personal pollution while taking samples from an outfall from ICI's factory in Billingham, UK. Scientific research into the state of the North Sea has always been part of Greenpeace's effort to cure the sea's sickness.

Fleet review
The MV *Greenpeace* flanked by the *Moby Dick* and the *Beluga* (above) gathered in Rotterdam harbour in March 1989. The three vessels then set out for separate series of actions around the North Sea as part of Greenpeace's campaign leading up to the third North Sea conference at The Hague in March 1990.

Among other research, Greenpeace commissioned experts to check how each country was meeting the commitments to reduce pollution that had been made at the second conference. The preliminary reports, which demonstrated that those objectives were not being met, were released in November 1989, with the intention of forcing the governments concerned to take the action that Greenpeace believed necessary to begin the process of cleaning up the sea – or to face the embarrassment of having their failure to give practical substance to the precautionary principle made public.

Although the UK and France had insisted that no information should be passed to any environmentalist group during the preparations for the third conference, the Danish parliament required that its government keep its affairs public, and from July 1989 the Danish papers on all the issues that the North Sea nations were discussing came automatically into the activists' hands. And, before long, packages containing conference documents arrived – anonymously – at the Greenpeace offices from most of the capitals of Europe.

For Greenpeace, this was an ideal vantage point from which to develop a political strategy. Once more, the first aim was to continue to isolate and expose the United Kingdom, still the biggest source of opposition to change. But the organization was putting pressure on other countries, too. The Nordic and Wadden Sea states had been highly critical of the UK, but Greenpeace were able to point to the many problems that existed in those countries and charge them with hypocrisy. Greenpeace also urged the Netherlands to put pressure on Belgium and France, who were causing massive pollution of the Scheldt estuary, which the three countries shared. This also helped to ensure that the UK would again be left stranded as the least co-operative and most destructive of the North Sea states.

In some ways fate was not kind. The Dutch government was originally due to hold a general election immediately after the third North Sea conference it was hosting, and could be expected to do its utmost to make sure that the conference

No time to waste
One of Greenpeace's slogans for their campaign leading up to the third conference is planted by the side of a waste outlet, belonging to the BASF factory, in Antwerp harbour. Activists have partially blocked the pipe to reveal the nature of its contents.

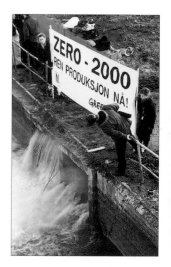

An end to all dumping
An action at the Peterson pulp and paper factory in Norway near the border with Sweden, in 1990, calling for a complete cessation of waste dumping by the year 2000.

was a dramatic success. But the government fell early in 1989 – ironically over how the protection of the environment was to be funded – and was politically inactive at a crucial time. The Swedish government was also forced into elections in spring 1989, with similar effects. And in the months before the conference, the dismantling of the Berlin Wall, and the prospect of reunification, meant that in Germany, at least, the North Sea was all but forgotten.

By the time the conference opened, however, the UK government had put itself out on a limb. It had vowed to continue sewage sludge dumping until 1998, and had renewed permits to continue dumping toxic industrial waste, despite signing the second North Sea conference's commitment to end dumping by the end of 1989, and it proposed to bury radioactive waste under the seabed off Dounreay in Scotland. The effect of all this was aptly summed up by the Dutch minister for the environment, who had also chaired the conference. There had, she said, been two major problems at the meeting: the failure to reduce nutrients – and the United Kingdom, who had consistently blocked progress across the board.

The heart of the matter

The ultimate purpose of Greenpeace's political strategy was to demonstrate the difficulties all countries were facing in fulfilling the agreement made at the second conference. If people could be made to see how desperate matters had become, they would surely then ask what could be done.

Answering that question involved recognizing that the only hope for the North Sea's recovery lay in stopping *all* discharges and dumping of contaminating waste into the sea as soon as possible. And so, in the middle of 1988, was born the Greenpeace slogan 'Zero 2000' – the call to stop all pollution entering the North Sea by the end of the century at the latest.

Such a radical demand naturally brought a further question: how was industrial waste to be disposed of? Greenpeace's answer was to promote the practice of 'clean production', which would drastically reduce the amount of waste produced in the first place. Greenpeace pressed the North Sea states, both before the 1990 conference and in the five-minute address they were permitted to give before the assembled ministers, to plan a programme of progressive reductions in pollution that would end the daily contamination of the sea by the year 2000. This, they said, should go hand-in-hand with a massive investment in all aspects of clean production. And Greenpeace demanded that the conference's resolutions should be made legally binding on all the North Sea states.

Greenpeace did not succeed in getting that essential change in emphasis included in the Final Declaration, although the Nordic states did make that commitment, and the European Commissioner for the environment, Carlo Ripa di Meana, in his opening address to the conference, promoted clean production as a policy that the North Sea states should adopt. The Declaration does, however, commit the North Sea states to further development of low-waste and non-hazardous products, to reduce discharges into the sea to levels 'harmless to man or nature' by the year 2000, and to address the question of making the Declaration legally binding. And three weeks after the conference, the European Parliament passed a resolution calling for zero discharges into the North Sea by the year 2000. Greenpeace's campaign was at least now firmly on the political agenda for the North Sea states.

In 1980, Greenpeace were inexplicable extremists in most people's eyes. Today, few people can claim not to know that the environment is in a state of crisis. Greenpeace have become the sharp, if still desperately needed, end of a growing popular consciousness that humanity must once more be the stewards, not the exploiters and polluters, of an increasingly fragile planet. As long as that idea survives, there is a glimmer of hope for the North Sea.

Stemming the tide
A Greenpeace activist battles to plug a pipeline discharging gypsum sludge from the Kemira fertilizer factory at Rotterdam in June 1989. The plant pours 1.2 million tonnes of sludge into the North Sea every year, contaminating the water with nutrients, heavy metals and radioactive substances.

Faces of the future
Dutch children show their concern for their heritage. Greenpeace became famous for their direct actions, but have always backed these with scientific research and political and industrial lobbying. In latter years this behind-the-scenes activity has included an educational programme, helping to make people aware as never before in an industrial society that their own interests – even their own survival – depend ultimately on the health and infinite variety of the natural world.

PART 4

We all want the North Sea to survive. Although we are causing many problems in the sea, we are also beginning to find solutions to them. Slowly, we are learning to be far more cautious in our expectations of what the North Sea can bear. We are beginning to realize that the sea cannot absorb the pollution we create, that we need to be far more careful about how many fish we take from the sea and, above all, that wildlife – not people – should have first call on the North Sea.

But there are limits to what individuals or even a single country can do. One nation alone cannot protect the North Sea, even if it stops all pollution, never catches another fish and vows never to reclaim another square centimetre of coast. The North Sea requires co-ordinated international action to save it.

The North Sea ministers' conference is intended to provide that co-ordination. It has formally recognized the need for a cautious approach in our attitude to the North Sea, and has taken the first faltering steps towards reducing the damage we cause. But it has yet to take the holistic view and the long-term measures that are needed to turn what is a precautionary principle into practical reality. On this the future of the North Sea depends.

The tide reveals all
The incoming tide brings with it a mass of foam from an algal bloom. During the 1980s, algal blooms, diseased fish, breeding difficulties among seabirds, the collapse of fish stocks, the loss of coastal wetlands, and the washed-up corpses of seabirds and seals formed an unmistakable distress signal from the North Sea. Politicians at first denied that there was anything wrong with the sea, but they were overwhelmed by the mass of evidence, and their attitudes slowly began to change.

POLITICS
THE TIDE MUST TURN

ANYONE CONTEMPLATING the North Sea today could be forgiven for imagining that there had been no restraints whatever on those who caused the ecological disaster that it has become. But, in fact, for some decades it has been impossible to do anything in the North Sea without a licence. The dumping of halogenated waste, ocean incineration, the use of oil-based muds and anti-fouling paints, and the plundering of fish stocks in the North Sea may be mistakes, but they were, or are, all regulated mistakes.

Today, governments recognize that environmental problems know no borders, and regulation is largely international. But the regulatory bodies are frequently uncoordinated, their responsibilities sometimes overlap, and they largely ignore some crucial aspects of the sea, such as the fate of its wildlife.

The politics of corruption

Some regulatory organizations concerned with the North Sea have worldwide responsibilities. The most important of these are the United Nations' International Maritime Organization's conventions on marine pollution (MARPOL), on oil pollution (OILPOL) and on sea dumping (the London Dumping Convention).

To tackle local problems more efficiently, the governments around the northeast Atlantic created two sister commissions: the Oslo Commission (OSCOM), in 1972, concerned with dumping, and the Paris Commission (PARCOM), in 1974, which regulates pollution from land-based sources. These negotiated legally binding agreements on the amounts of various chemicals that could be dumped or discharged. But they were bound to consider the impact on wildlife only when it qualified as a 'natural resource', and this meant ensuring only that the human ability to catch or eat fish was not impaired.

The search for proof
A Dutch research ship monitors a natural algal bloom in the North Sea. Despite many years of research there are still huge gaps in our knowledge of the North Sea, and of the impact we have upon it. Further investigation is essential, but we should not wait for scientific proof that an action damages the environment – the suspicion that it might should be a sufficient spur to action.

No harmful effect
The UK continues to dump colliery spoil in the North Sea (left), arguing that the waste is chemically inert and therefore not harmful to the environment. But the volume and physical nature of the waste mean that it smothers the shore and seabed, destroying all life. For many years regulators have attempted to assess which substances can safely be discharged into the sea, but damage continues to occur. The solution is to avoid producing the waste in the first place, or if this is unavoidable, to store it in such a way that it is isolated from the wider environment.

There was a deeper flaw, too. The regulators took it for granted that the sea could absorb a certain amount of waste each year; that it had an 'assimilative capacity' for pollution. And to cut down the task of assessing the effects of tens of thousands of chemicals on thousands of marine species, they concentrated on the effects of what they thought were the most damaging 'persistent, toxic and bioaccumulative' substances. But research eventually showed the authorities that pollution entering the sea was not disappearing, that they could not even assess the damaging properties of a substance, and that, in any case, different problems were constantly being created, sometimes by substances such as nutrients that had not previously been recognized as pollutants.

The wrong way round

In reality, the problems of establishing which substances were harmful meant that no action was ever taken against a pollutant until its destructive effects appeared. Moreover, those who were concerned about the environment were being asked to show which particular substance was responsible for a particular problem – although the regulators themselves had found it impossible to make such precise correlations. Meanwhile, the stock reply of industry to criticism was that it adhered strictly to international guidelines or regulations.

Although the regulators and industry may have convinced each other that they were doing the best possible job, evidence to the contrary began to present itself in a procession of dead birds, diseased fish and algal blooms.

THE MEETING POINT

The threatened state of the sea had become obvious to almost everyone by the early 1980s. The practical, political result was the North Sea conference. At last, the entire North Sea ecosystem would be discussed in one forum and, because it was a meeting of environmental ministers, it would have great political power.

The first North Sea conference took place at Bremen, in West Germany, in autumn 1984; its main achievement was to have met, and agreed some common ground, at all. The second conference, held in London in November 1987, was the scene of some remarkable political manoeuvring and two remarkable results. The eight nations agreed to adopt the 'precautionary principle' as their guiding light in protecting the North Sea and, as a result, to reduce discharges of what they believed to be especially toxic substances, and nutrients, to 50 percent of their 1985 levels by 1995. The second conference also agreed to end dumping of industrial waste by the end of 1989, to stop ocean incineration by the end of 1994, to reduce the disposal of rubbish, oil and chemicals from ships at sea, to reduce the amount of oil being discharged in oil-based muds from drilling platforms, and to set up a Scientific Task Force to harmonize research into pollution of the North Sea.

The third North Sea conference

The seal epidemic and a number of huge algal blooms in 1988 heightened public awareness of the state of the environment, and there were great expectations that positive steps to help it would be taken at the third conference, held at The Hague in March 1990.

The North Sea states added atmospheric pollution to those subject to a 50 percent reduction by 1995, and decided to reduce pollution from some substances, including dioxins, by at least 70 percent by 1995 – although the wording of the final declaration shows that some countries were leaving room to back out of these commitments even as they made them. Ocean incineration was now to end by 1992. And, having adopted the precautionary principle, it was only logical that they should commit themselves to recovering, or making

Fish disease and pollution
In the early 1980s evidence began to accumulate that fish living in heavily polluted areas of the North Sea suffered a high incidence of disease. This discovery was one important factor in persuading politicians that the time had come for action to save the North Sea.

THE PRECAUTIONARY PRINCIPLE
Applying the precautionary principle means giving the environment the benefit of the doubt if there is uncertainty about the effect that an action may have on it. The principle has its origins in the German concept of the *Vorsorgeprinzip*, developed in the 1970s and expounded specifically for the North Sea in a 1980 government report, as a result of serious levels of pollution in the German Bight.

The decision, taken at the second North Sea conference in 1987, to apply the precautionary principle came in the wake of the problems caused by the previous 'permissive principle', which allowed chemicals, for example, to be dumped in the sea unless there was proof that they were damaging. This had its roots in a view of the world as a huge place in which human actions had little impact. That view was false, but it was often difficult to show convincing evidence of harm until after the event.

Although the North Sea states now promote the precautionary principle elsewhere, they themselves have been slow to realize its full implications.

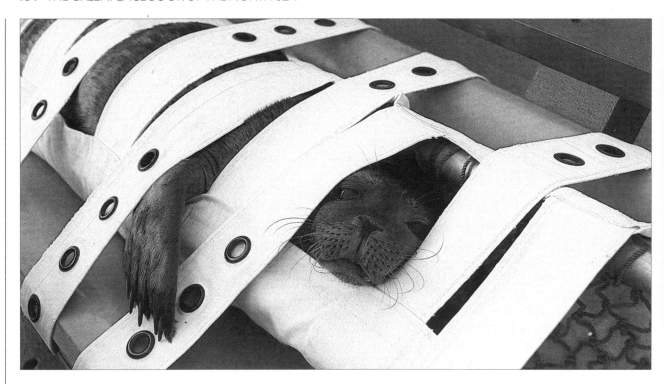

harmless, sunken hazardous cargoes such as pesticides or toxic waste, with the polluter or potential polluter paying the bill.

Seven states declared that the North Sea and its seabed were not suitable dump sites for radioactive waste – while the United Kingdom, the only country actually contemplating such a move, refused to accept this agreement. The UK agreed to stop dumping industrial waste by 1993, having already failed to meet the 1989 deadline set at the second conference.

Progress since 1987 in reducing nutrients by 50 percent had been slow, and the third conference, and its final declaration did little to help. Only the larger sewage treatment plants had to meet specific, lower levels of nitrogen and phosphorus discharged, and there was no agreement to reduce levels of atmospheric nitrogen. The UK promised to stop sewage sludge dumping – by the end of 1998.

Seal plague
The epidemic that affected North Sea harbour seals in 1988 demonstrates how the precautionary principle should operate. A good scientific case can be made that pollution played a role in the epidemic, even though it may be very difficult to prove. Politicians should therefore act to eliminate the substances that caused the pollution, without waiting for final scientific proof, one way or another, that may never come.

What have the conferences achieved?

The adoption of the precautionary principle at the second conference has undeniably become a landmark among international environmental agreements. The precautionary principle has since been adopted by the United Nations' Environment Programme, PARCOM, the Barcelona Convention for the Mediterranean, and the Nordic Council. And the agreement to go for 50 percent reductions of nutrients and persistent, toxic and bioaccumulative substances cut through years of fruitless argument about the assimilative capacity of the sea.

The third conference committed the North Sea states to make its decisions legally binding through national or EEC regulations, ending the claim that it was merely a 'political agreement' and that a minister's word had no real force.

But how real is the commitment to action? Even firm targets agreed at the second conference were never met – dumping at sea, for instance, did not end in 1989. The most significant achievement of the second conference was to adopt the precautionary principle but, two years later, only in West Germany, which had originated the concept back in the 1970s, had it become the foundation of environmental practice – and even there the practical results were questionable.

In 1987, the second conference called for 'urgent and drastic' action in reducing pollution by 50 percent. But in 1989, a set of independent reports from leading experts in each country was not reassuring. They found that many countries had only the most general of plans, if any, of how they would reduce persistent substances by 50 percent, and that the plans to reduce nutrients were in disarray. These findings were later confirmed by the official implementation report prepared for the third conference, which was unable to review a full set of plans for persistent substances, and candidly admitted that there were problems with the reduction of nutrients; some countries doubted that even 30 percent reductions were realistic.

The North Sea states will continue to fail in meeting their own objectives as long as they fail to work out the full implications of the precautionary principle.

TAKING PRECAUTIONS SERIOUSLY

Persistent, toxic and bioaccumulative substances are, by definition, persistent. The North Sea's contaminated estuaries and coastal waters show clearly that it cannot handle any further pollution. And these substances form only one kind of pollution.

This is where clean production comes in, with its emphasis on waste avoidance. This requires action by industry, consumers – and governments. The main problem is that the regulators are still fostering the belief that it is possible to find a 'safe' level of pollution. Until decision makers state clearly that priority must be given to avoiding the creation of waste, industry will not fundamentally rethink the way it operates.

Some waste, an ever diminishing quantity, will continue to be created in clean production. To cope with this, the regulators will first have to produce a list of permitted waste substances that have been shown, as far as is possible, to have no harmful effects on the environment. This reverses the current practice of listing substances only once they turn out to be damaging. But the list must be as short as possible: the fewer substances there are on it, the less chance there will be that one of them will prove to be harmful after all (PCBs, DDT, CFCs and oil-based muds are all substances that were certified 'safe' a few years ago). Industry must then ensure that its waste products conform to this single 'reverse' list. This policy can work only in a world that has sharply reduced the amount and diversity of its waste, through clean production. Such waste should be stored in such a way that it could not pollute the natural world, and where it could be retrieved.

A plan for action

Such a change to clean production requires long-term planning, clear goals and interim targets. The goal should be for North Sea pollution to be reduced to zero by the end of the 1990s. The North Sea conference in association with OSCOM, PARCOM and the EEC are the most appropriate bodies to oversee such a change. A programme for zero emissions to the North Sea by 2000 would call for:
- Clean production methods that already exist to be introduced in all North Sea states by specified dates before the North Sea conference set for 1995.
- A systematic plan, including target dates for all sectors of industry and agriculture, to develop fresh techniques for clean production as the central means to protect the North Sea during the 1990s.
- Reverse listing, to be applied to all waste that is generated.
- The setting up of an international network of institutes to establish clean production methods and to assist industry to apply them.

The plan should emphasize 2000 as the latest possible date and should ensure that in the late 1990s the last few percent of current discharges will stop. Unless this is done, some industries might put off research and practical action until the last possible moment. Once it is clear that clean production is the key to halting the production of waste, industry can begin to direct its research accordingly.

From sea to land
When ocean incineration ends in 1992, governments expect that hazardous waste will simply be burned on land. In practice, politicians still emphasize alternative methods of disposing of waste, rather than making waste avoidance the prime objective.

Trading in hazards
One of the worst aspects of the waste disposal industry is the international trade it encourages. Those who create hazardous waste can simply ship it to the countries that have the weakest regulations concerning disposal. If the practice of clean production were to be adopted on a broad scale, no toxic waste would be produced at all – and the risk to the environment would be drastically reduced.

PATHWAYS TO CHANGE

The problems that beset the North Sea stem mainly from six areas of human activity: the destruction of wildlife and natural habitats, shipping, pollution (from both industry and agriculture), the extraction of oil and gas, fishing, and military activities.

Public pressure (1) in response to an abuse of the environment leads to action by politicians (2–12). The problem may be so great or its solution so obvious that direct action is taken. Sometimes there is a long drawn-out controversy about the issue, involving non-governmental organizations – such as voluntary organizations, conservation and research bodies (13), scientific research (14), industry (15), legal issues (16) – even the governments of states beyond the North Sea (17) and international government bodies, such as the United Nations and its agencies (18). This public debate takes place, too, within a host of international, specialist regulatory bodies (19–42), which then make recommendations to the governments of the eight North Sea states (5–12).

Drawing on this wealth of often conflicting information, politicians make the final decisions on action – either nationally or co-operating in the European Community (2), the Nordic Council (3) or the three Wadden Sea states (4). The most influential form of co-operation has become the International Conference on the Protection of the North Sea (43).

Many of the issues are so complex that the politicians may refer back to the specialist regulatory bodies (19–42), on whose recommendations they will then make their decisions. These have an effect on the environment – whether for good or bad – and the whole cycle starts again.

THE REGULATORY AUTHORITIES
The international authorities that regulate activities in the North Sea tend to concentrate on specific issues:

Wildlife
The International Union for the Conservation of Nature (IUCN 19) is a worldwide body that includes governmental and non-governmental organizations. It covers two often irreconcilable issues: the protection of wildlife, and its exploitation. The IUCN's main influence in the North Sea has been through its sponsorship of the Ramsar Convention (20), which obliges North Sea states to protect coastal wetlands of global importance. The Bonn Convention (21) concentrates on the protection of migratory animals. The Bern Convention (22) was set up by the Council of Europe, and provides basic levels of protection for wildlife by safeguarding habitats. Both the Bern and Bonn Conventions tend to focus on land-based problems. The Wadden Sea states (4) worked towards the Bonn Agreement on the Conservation of Seals in the Wadden Sea (23).

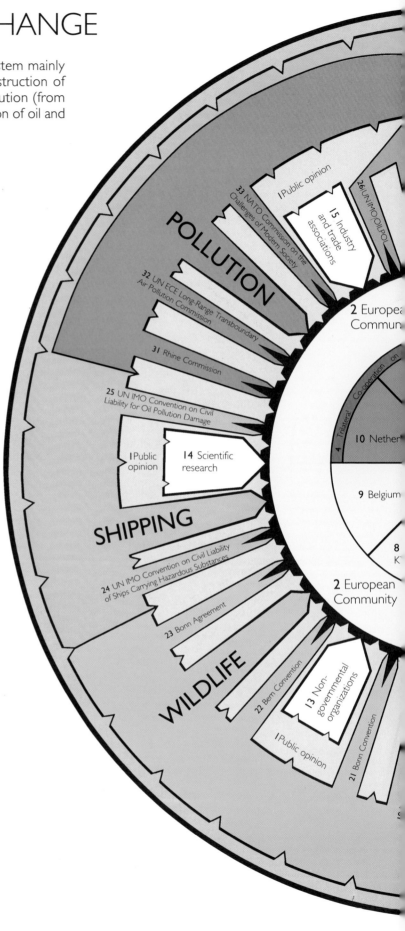

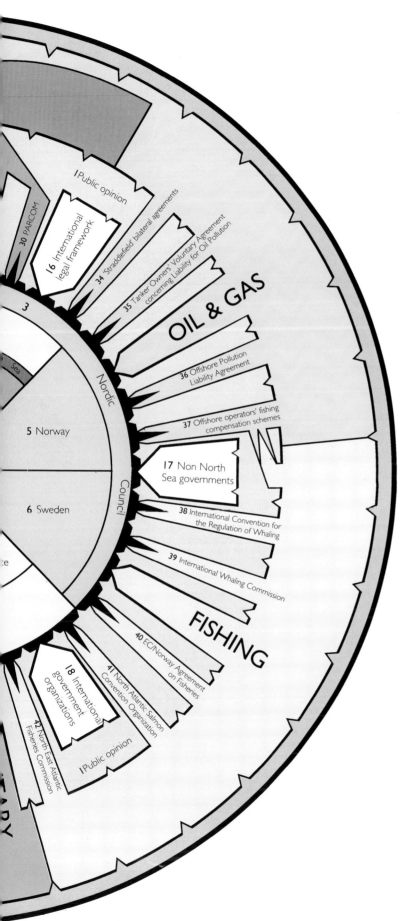

Shipping
Shipping is regulated globally by the United Nations' International Maritime Organization (IMO). Its instruments include a proposed Convention on Civil Liability of Ships Carrying Hazardous Substances (**24**), a similar convention on liability for oil pollution damage (**25**), regulations concerning oil pollution itself (OILPOL **26**), marine pollution (MARPOL **27**) and dumping at sea (the London Dumping Convention **28**).

Pollution
Two of the most important bodies governing the North Sea are the Oslo Commission (OSCOM **29**), which regulates dumping in the northeast Atlantic, and the Paris Commission (PARCOM **30**), which regulates land-based pollution in the same area. The Rhine Commission (**31**) regulates pollution entering the river – a major source of pollution in the North Sea. The UN Economic Commission for Europe (**32**) deals with international atmospheric pollution. NATO's Committee on the Challenges of Modern Society (**33**) has produced detailed reports on pollution problems, but it is not linked to any decision-making process.

Oil and Gas
OSCOM and PARCOM are the main international regulators of pollution from the offshore and onshore oil and gas industries. Issues, including pollution, that affect oil and gas fields extending across two states' territorial waters may be dealt with by bilateral 'Straddlefield' commissions (**34**). MARPOL and the LDC also have some involvement in the offshore industry. There are limited, voluntary liability agreements for tanker owners (**35**), working in conjunction with the IMO's convention (**25**), and for offshore operators (**36, 37**).

Fishing
Two organizations, the International Convention for the Regulation of Whaling (**38**) and the International Whaling Commission (**39**), restrict the hunting of all great whales. The Agreement on Fisheries (**40**), between the EC and Norway, along with the Common Fisheries policy of the EC (**2**), gives comprehensive coverage of fisheries issues and largely displaces earlier organizations. However, the North Atlantic Salmon Convention Organization (**41**) and the North East Atlantic Fisheries Commission (**42**) still play minor roles.

Military
There are no international bodies that regulate damage caused by military activities in the North Sea. These issues are dealt with at national level, if at all.

Economics and environment
The inside of a tannery in northeast England. One fear that many companies have is that measures to ensure a better environment will be expensive, but this is not necessarily the case. The leather industry, among others, once used oil from the sperm whale, driving the species to the verge of extinction. Only when the oil seemed likely to be banned was an alternative sought – and a cheaper substance found.

The safety of a national park
A new drainage channel in the Nordstrander Bucht, a reclaimed site in a German national park in the Wadden Sea. National parks were set up to preserve the natural landscape – yet unless there is the political will to make them inviolable, habitats in national parks will simply be destroyed more slowly than elsewhere.

Obstacles to change

One traditional obstacle to environmental protection programmes has been the fear that they will increase production costs and cause unemployment. Industrialists argue that the cost of a filter system, for instance, grows exponentially as one tries to remove ever smaller amounts of the pollutant.

But clean production shifts the emphasis away from technological fixes at the end of a pipe or smokestack. For instance, using benign raw materials instead of hazardous ones removes the entire environmental problem instantaneously, and results in huge savings for a company on the cost of a filter system.

Not just companies but a whole region or country can benefit from clean production. Bodies such as the Organization for Economic Co-operation and Development have concluded that since the 1960s protective measures either have not affected national economies or have even increased national wealth by 1–2 percent. And this was when regulators concentrated on expensive end-of-pipe or smokestack controls. Employment actually increased, as a new sector of industry grew, devoted to preventing pollution, and other industries and the health of employees also benefited.

But often pollution has continued not because solutions do not exist, nor because they cost too much, but simply through inertia. One piece of research concluded that in 60 percent of cases, bureaucratic resistance, unwillingness to change, conservatism, piecemeal legislation, poor publicity, and misunderstandings stood in the way of progress to clean production.

We are all involved

Everyone can play a part in breaking down resistance to change. The consumer can choose between products that are harmful and those that are not. The engineer can look for ways of eliminating waste, or carry on as before. The manager can make an immediate investment in developing clean production, or keep the old plant running as long as possible. Governments can change economic structures to help to make clean production work, or look on passively. Unless we all act, the environment will continue to suffer – and so will we.

WILDLIFE AND FISHERIES

While the regulatory authorities have concentrated on pollution, they have paid scant attention to protecting wildlife, and the welfare of the North Sea's plants and

animals is low on the political agenda. It is extraordinary how little emphasis is given to the interests of wildlife by bodies such as OSCOM and PARCOM.

The 1971 Ramsar Convention is supposed to preserve wetland habitats of global importance in the North Sea and elsewhere, but some governments have been very slow to co-operate in protecting them. The Bern Convention's task is to protect animals by preserving their habitats, while the Bonn Convention tries to ensure that each link in the chain of habitats for migratory species is safe. Bern and Bonn have been criticized both for their general weakness and for not paying sufficient attention to marine life. Even though many marine species and birds range freely across the North Sea, there are few internationally agreed protective measures, and no recognition that the ecosystem as a whole has to be protected.

The 1988 Bonn Agreement to unite the efforts of the three Wadden Sea states to protect harbour seals, and moves to extend this policy to the North Sea as a whole, mark an encouraging change in attitude. So, too, does the inclusion of issues concerning wildlife and fisheries at the third North Sea conference, although the only firm agreement reached there – to investigate ways of reducing the numbers of small cetaceans accidentally caught by the fishing industry – was not as far-reaching as many had hoped it would be.

Changing times
Fishing is a fiercely independent industry, much of it made up of very small family businesses, and those who work in it do not find it easy to transfer their skills to another trade. As fish stocks dwindle, the industry has to buy ever more expensive equipment, while employing fewer people and catching fewer fish. The politicians now need to find a way of reversing this trend.

Hunters of the sea

Fishing regulation has a long but chequered history. Fishing in the North Sea is now largely controlled through the EEC's Common Fishing Policy (CFP), along with the Joint Agreement between the EEC and Norway. The CFP came into operation in 1983. Each year, scientists gather information on the status of fish stocks, from which they recommend a Total Allowable Catch (TAC) for each species. Politicians then share the allocations among the different countries, adjusting them to produce an 'agreed TAC'. During the year, statistics are gathered, so that the approximate size of the actual catch is also known. The politicians usually increase the size of the recommendation, and the actual catch usually exceeds the agreed catch, sometimes by more than 50 percent. For this reason, the North Sea's fish stocks are still in dire straits.

Then there is the question of beam trawling. By the mid 1980s, 309,000 square kilometres, the equivalent of over half the area of the North Sea, were being disturbed annually by beam trawling, which drags heavy chains across the seabed to raise fish there into the nets. The technique has changed the nature of the life on the sea floor in coastal areas and the southern North Sea, because it crushes shellfish, while small worms thrive in the constantly disturbed sediments.

Prohibiting fishing in certain areas could not only help the ecosystem to recover, but also result in increased catches for the whole North Sea, if stocks recover in protected areas and migrate to others. Indeed, such a strategy could be useful in surface and mid-water fisheries as well as for beam trawling.

Just as with pollution regulation, the most peculiar aspect of the controls on fishing is that, although 25 percent of the North Sea's fish are caught each year, no one considers the impact on wildlife. The vast number of fish taken from the North Sea throughout the century has probably resulted in a decline in its populations of porpoise and dolphin but, as always, this is very difficult to prove. The same is true of the relationship between industrial fisheries and the decline of some seabird colonies. A precautionary approach dictates that those who want to remove the sandeels should prove that they are not harming the birds that feed on them – but any rational policy would recognize that the two interests coincide.

Somewhere to live

Not all of wildlife's problems in the North Sea come from overfishing. The loss of habitat is another huge problem, and may be subtle. For instance, cetaceans are sensitive to noise transmitted through the water, and oil and gas prospecting, along with ship and power boat engines, have made the North Sea a very noisy place under water. Gravel and sand extraction are licensed to avoid areas important to fish, but the operators must take care to leave further stretches of the seabed undisturbed so that the fish have room to move should their distribution change, either as their populations recover from decades of overfishing, or as a result of changes in climate. Habitats have also been destroyed in the wetlands, mud flats and estuaries. The many legitimate human users of these coastal areas must now realize that the marine mammals, birds and fish that depend upon them have nowhere else to go – and must now have first call upon them.

THE FUTURE

The next North Sea conference, planned for 1993, will be a joint meeting between environmental and agricultural ministers. It should be an ideal opportunity to bring wildlife and fisheries fully into the agreements to protect the North Sea, as many of the responsibilities for them lie with agriculture ministries. OSCOM and PARCOM, now showing new signs of life, will have their first ever ministerial conference in Paris in 1992, where the practical steps to implement the precautionary principle ought to be taken, and extended to the northeast Atlantic as a whole. The EEC Environmental Commissioner's support for clean production at the third North Sea conference, along with that of the European Parliament, may bring positive results, while a new EEC Directive on ecological protection could also help wildlife in the North Sea.

It is not that solutions do not exist for the problems of the North Sea – simply that they are not being acted on. The North Sea is threatened today because of short-term, piecemeal approaches to its protection. Environmental policy is at present confused, caught between the old permissiveness and the precautionary principle. It is up to all of us to make sure that the precautionary principle becomes a practical reality.

Choose or lose
A rorqual whale cast up on the Belgian coast. The sight of a stranded whale never fails to draw the curious, astonished that something so large and strange exists in a sea that is not now regarded as exotic. Whales and dolphins are indeed far scarcer in the North Sea than they were a century ago – a decline that is due to hunting, overfishing, and pollution. When the cetaceans have returned to their former abundance we will know that we have succeeded in restoring the North Sea.

INDEX

Page numbers in *italic* refer to captions.
Those in **bold** indicate main entries.

A

accidents at sea, **112-116**
acid rain, Greenpeace campaign, *121*
agriculture, effects on North Sea, 82-85
aldrin, 90
algae, 55, 58, 59, 62, 75, 80-86, *80*, *81*, *82*, *84*, 86, 133, *140*
　cause of deoxygenation, 82
　toxins in, 80, **83**
Altenbruch, Germany, 63, 65
aluminium, discharge of, *91*
ammonia, 84
Amoco Cadiz, grounding, 113, *113*
anchovy, 67
anemone, 28, 29, *32*, 46, 54, *54*, 55
　beadlet, 58, *58*
angler fish, 29, *106*
Anna Broere, sinking, 113
anti-fouling liquids, 92, 93, *106*, *132*
Antwerp, *122*, 126
　Greenpeace actions, *120*, *122*, *128*
　port development, 65
APC, fertilizer manufacturer, 122
Ark Royal, Greenpeace action, *120*
arsenic, 90
Atlantic Ocean, 14, 18, 28
atmosphere, pollution of, 87-89
　agreement to reduce, 133
　dioxins in, 100
　from industry, 96
　from power stations, 88
　from shipping, 115
　from vehicle exhausts, 88
　heavy metals in, 93, *93*
　nutrients in sea from, 87-89
　pollution routes to sea, 92
auk, 40, 41, 42, *111*
　mercury levels in, 94

B

bacteria, 45
　use of in sewage treatment, 86
Baltic Sea, 14, 18
Barcelona Convention for Mediterranean, 134
barnacle, 29, *34*, 54, 58
BASF, Antwerp, factory waste outlet, *128*
Bayer chemical company, 120-122, *121*, *124*
beaches, destruction of, *66*, 67
　litter from shipping, 114
Belgium, coast of, 67
　dumping, Greenpeace target, 122
　fishing in North Sea, 71
　import ban for ocean incineration, 126
　ocean incineration of HHCs, 100
　pollution of Scheldt estuary, 128
　ports, handling of shipping waste, 114
　salvage of uranium hexafluoride, *112*
　sewage effluent, discharge of, 87
Beluga, Greenpeace ship, 123-124, *127*
benthic zone, 45, 82
Bern Convention, 136-137, 139
bib, 58
Billingham, UK, Greenpeace action, *127*
bioaccumulation, *91*
bioaccumulative substances, 133, 135
　agreement for reduction of, 134
birds, 28, 40, 56, 59, 77, 83, 139
　and oil pollution, *111*, 113
　danger from litter, 114
　displaced by development, 65
　PCB levels in, 95
Blakeney Marsh, East Anglia, 62
blenny, 55
Bonn Agreement, 136-137, 139
Bonn Convention, 136-137, 139

Brandenburg, collision, 112
Bravo oil platform, blow-out, 110, *110*
Britanny, beaches, oil contamination, 113, *113*
brittle-star, 29, 45, 54, *54*, 55

C

cadmium, 90, 92, 120, 122
　conversion by benthic life, 93
　effects on plankton, 101
Cap de la Hague, radioactive waste, *120*
car *see* motor vehicle
carbon, 29, 62, 94
carbon cycle, 94
carbon dioxide, 29, 67
　concentration in seawater, *91*
carbon monoxide, 88
cargoes, hazardous, international agreement to recover, 133
　loss or spillage, *112*, 113-115, *114*
　regulation of, *114*, 115
catalytic converter, 88
catalytic reduction process, 89
cetaceans, 43, 77, 139-140, *140*
　effects of pollution on, 77
CFCs, 95
Channel Islands, 113
chemical industry, 101-103
chemical pollution, agricultural, 90, *90*, 96
　anti-fouling agents, 92, 93, *106*, 132
　assessing effects on marine species, 133
　discharge of ships' waste, 114
　dumping waste, 120-121
　effects on mammals, *91*
　from industry, 89, 103
　ocean incineration *99*, 125
chemical weapons, 116
Cherbourg, nuclear fuel at, *102*
chromium, 90, 92, 94
Chrysochromulina polylepis, 80, *80*, *81*, 82, 83
Claymore oil field, spillage, 110
clean production, 101-103, *101*, 135, *135*, 138
　EEC support for, 140
　promoted by Greenpeace, 126, 129
climate, 14, 18
　and global warming, 67
coal, formation of, *20*
　use in power stations, 87, 89
coal mine, pollution from, *91*, *132*
coastal erosion, 63, *66*, 68-69
coastal waters, 40, 42, 54, 56, 82
　pollution in, 108,135
coccolithophore, *81*
cockles, 29, *34*, 51, 68
cod, 29, 33, 37, *38*, 40, *40*, 41, 45, 58, 67, *73*, 76, 80
　fishery, 70-71, 73
　methyl mercury in, 93
Cofaz, fertilizer manufacturer, 122
comb jelly, 30, *31*
　effects of cadmium on, 101
copepod, *29*, 30, *30*, *32*, 33, 36, 52
　effects of pollution on, 101, 108
copper, 90, 92
　biological function of, 92
coral, 28, 54, *54*
cormorant, 56, 58
crab, 28, 29, *29*, 31, *32*, 45, 45, 52, 55, 68, 75, 78, 80
　burrowing, 29; edible, *52*, 55; hermit, *50*, 55; mitten, 55; porcelain, 31, 55; shore, *53*, 58; spider, *52*, 55; swimming, *52*, 55
crawfish, *32*
crustacean, 30, 32, *38*, 45, *45*, 51, 52, 57, 58, 59
curlew, 59, *62*
cuttlefish, 42, 50

D

dab, 123
　HHC accumulation in, 100
dams, 68
Dana Optima, loss of toxic cargo, 113
DDT, 95, 98-99
dead man's fingers, 54, *54*
Delta Plan, Netherlands, 68-69, *68*
Denmark, plan to reduce nutrient pollution, 85
　and environmental groups, 128
　fishing in North Sea, 71, 77
　gas and oil platforms, 104
　Greenpeace in, 120-121
　increase in nitrogen, 82
　increase in phytoplankton, **82**
　mud flats, *34*
deoxygenation, 82
diatom, 29, *29*, 33, 59, **82**
dieldrin, 90
diesel engines, exhaust emissions, 87-88
dinitrophenol, cargo lost at sea, 113
dinoflagellate, 29, *29*, 33
Dinophysis, 83
dioxins, 100, *100*, 109, 125-126
　international agreement to reduce, 133
dolphin, 28, 36, 42, 43, 77, 83, 139, *140*
　bottlenose, 42, 64, *95*; common, 42, *42*; Risso's, 42; white-beaked, 42
　effect of survey explosions on, 105
　HHC levels in, 95
　mercury levels in, 94
　and fisheries, **77**
Dounreay, Scotland, radioactive waste, 120, 129
Dover Straits, 14, 43
　atmospheric pollution in, *115*
　shipping in, 112
Downing Street, London, Greenpeace action, *127*
dredging, and heavy metal pollution, 93
　contaminated sludge storage, *94*
duck, 56, 58, 62-63, 67, 68
dumping, 120-121, *132*, *132*
　agreements to end, 129, 133
　radioactive waste, 129, 133
　toxic waste, licences, 121, 129
dunlin, 59, *59*, 65

E

East Germany, use of lindane, 98
Eastern Scheldt, Netherlands, damming of, 68, *68*
fisheries in, 68
ecology of the sea, **34-35**, 54
EDC tars, disposal of, 99
EEC, 135
　Common Fishing Policy (CFP), 139
　directive on catalytic converters, 88
　directive to end dumping, 121
eel, 29, 48, 68
eel grass, 29, 64
eider, 56, 58, 78
Elbe, river, 14, 69, 93
electricity, generation of, 89
endrin, 90
English Channel, 14, 28, 42, *42*, 54, 113-114
　currents, 18
　military exercises in, 117
　seals in, 38
Environmental Commission, for reduction of toxic substances and nutrients, 124
Environmental Protection Agency, US, 103, *103*
estuaries, 56, 67, *67*, 69, 82, 96
　destruction of habitats, 140
　Greenpeace analysis, 124
　heavy metal pollution, 93
　industrial developments in, 65

estuaries *(continued)*
　oil contamination, 108, *111*,113, *113*
　pollution in, 135
European Parliament, and clean production, 140
　zero discharges, 129
eutrophication *see* overfertilization

F

Fagbury Flats, England, development, 65
Falco, dump ship, Greenpeace protest, *121*
farming *see* agriculture
ferries, *112*, 113
　disposal of waste from, 114
ferrous sulphate, dumping, *121*
fertilizer, 84-86
filter-feeders, *32*, 45, *46*, 51
fish, 28, *32*, 33, 36, 37, 40, 42, 48, 55, 56, 58, 67, 83, *83*
　absorption of oil, *106*
　and ocean incineration emissions, 126
　disease and deformity, 95, 123, *133*
　effect of titanium dioxide on, 121
　numbers caught, 139
　PCB levels in, 95, 96
　territorial, *34*
fish farming, **75**
fish larvae, 29, 30, 31
fishing, **70-79**, 132, *139*
　EEC regulation of catches, 139
　effects on ecosystem, 76, 77
　fisheries management, 72-73, 79
　government actions, 72-73
　industrial fisheries, 64, 71, 74-75, 76, 77, 139
　overfishing, 79, *79*
　pelagic (surface waters), 70, 73
　predicting fish populations, 72-73
　protected areas, 139
　regulations, 139
　size of catch, 70, 74, 76
　target catches (TACs), 79
fishing fleets, 71
fishing methods, beam trawling, 71, *71*, 139
　echo location, 73
　fixed nets, 79
　otter trawling, 70, *71*
　purse seining, 71, *71*, 73
　seine netting, 71
fishing tackle, dangers to wildlife, 77, 114
fjords, algae in, 82
　atmospheric pollution in, *115*
　fish farming in, 75
　red tides in, 83
flat fish, 29, 33, 38, 45, 48, 51, *51*
　fishery, 73, 74
flounder, 51, 123
food web, 29, 30, *32*, 36, 40, 43
　and bioaccumulation, *91*
　and global warming, 67
　and oil pollution, *106*, 108
　benthic, 45
　methyl mercury in, 93
　monitoring pollution effects on, 101
　of herring, 33
　plankton, 36
　pollutants in, *91*
fossil fuels, burning of, 82, 84
　emissions from, 87, 89
France, agricultural lime extraction, 109
　coastal waters, increase in nitrogen, 82
　increase in phytoplankton, 82
　fishing in North Sea, 71
　Greenpeace in, 120-121
　gypsum sludge dumping, 122
　pollution of Scheldt estuary, 128
Frisian tidal flats, 63
fulmar, 56, *57*, 77

G

gannet, 28, *40*, 41, 56
gas, formation of, 20, *20*
    use in power stations, 87
gas and oil industry, **104-111**
    noise from, 140
    platforms in shipping lanes, 112
    See also oil
German Bight, cod fishery in, 73
    herring fishery in, 70
    increase in nitrogen, 82
Germany, agricultural lime extraction, 109
global warming, **67**, *67*
goby, 55, 58, 80
godwit, 59
Gonyaulax excavata, 83
goosander, 56, 58
goose, 56, 62-63, 67
    brent, 64
gravel, *16*, 18, 45, 51, 54, *63*
    mining in North Sea, 109, 112, 140
greenhouse gases, 67
Greenpeace, ship, *127*
Greenpeace, North Sea campaigns, **120-129**
    influence on North Sea conferences, 123-125, *123*, *125*, 127-129, *127*, *128*
    pressure on countries to implement agreements, 128-129
    raising public awareness, 120-121, 124-125, *127*, 129
    scientific research back up, 120-121, 123-124, 127-128
guillemot, 28, *40*, 40, 41, 56, 96, 110, *111*
    black, mercury levels in, 94
    breeding colony, 40
gull, 28, 56, 58, 67
gypsum sludge, dumping of, Greenpeace target, 122, *122*, 129
Gyrodinium aureolum, 83

H

habitats, loss of, **62-69**, *63*, *66*, *67*
haddock, 29, 45, 67, 71, 76, *76*
    decline in population, 74
    fishery, 70-71, 73, 74, 79
hagfish, 45, 48
Hague, The, sewage sludge, 87
halogenated hydrocarbon see HHC
halogens, 94
Hamburg, Greenpeace action, *120*, 121
Havergate Island, UK, bird reserve, *63*
heavy metals, *91*, 92-94, 109, *129*
    discharge into atmosphere, 93, *93*
    discharge into rivers and sea, 93, *93*, *94*
    dumping, 121
    effect on copepods, 101
    in Humber estuary, 96
    routes to sea, 93
herbicides, 63
    loss at sea, 113
Hermaness, puffin colony, 77
herring, 28, 31, 33, 36, 37, 38, 40, *40*, 41, 43, 48, 54, 67, 71, *72*, 74, 76, 96
    effect of gravel extraction on, 109
    effects of algae on, 80
    fishery, 70, *72*, 73, 76, 77
    food web, 33
    shoaling habit, *37*, 70
    Zuider Sea, 64
HHCs (halogenated hydrocarbons), 90-91, 94-95
    disposal of, 99
    effect on copepods, 101
    from ocean incineration, 98, *99*, 125
    from pulp and paper industry, *100*
    in pesticides, 98-99
    routes to sea, 98-99, 125
Højer sluice project, Wadden Sea, 63
Humber, river, 67
    industrial pollution in estuary, 96
hydrocarbons, 88, 94
    pollution from oil and gas industries, *106*, *108*
    polyaromatic, 88, *111*
hydrochloric acid, *125*

I

ICI, UK, outfall from factory, *127*
IJsselmeer, damming of, 64
incineration see ocean incineration
industrial fisheries see fishing
industry, and environmental protection programme, 132, 138
    changes for zero pollution emission, 135, 138
    draining of wetlands for, 62-63, 65, *65*
International Stomach Sampling Project, 76
iron, biological function of, 92
    discharge of, *91*

J K

jellyfish, 30, *30*
Joint Agreement, EEC and Norway, fishing, 136-137, 139
Kattegat, 14, 80
    chemical weapons in, 116
    currents in, 18
    increase in nitrogen, 82
    methyl mercury in cod from, 93
kelp, 29, 55, 58
Kemira, fertilizer factory, Greenpeace action, *129*
kittiwake, 56, 77
knot, 59, *59*, 65
Kronos-Titan, chemical company, 121-122, *121*

L

lamprey, 48
land reclamation, **62-66**, *64*, *65*
landfill sites, and heavy metal pollution, 93
Landskrona, Sweden, clean production in, *101*
lead, *88*, 90, 92, 120
    concentration in seawater, *91*
    from exhaust emissions, 93
Leverkusen Bridge, Greenpeace action, *124*
limpet, 33, 58, 80
lindane, 90, 95, 98
    cargo lost at sea, 113-114
ling, 29, 80
litter, danger to wildlife, 114
lobster, 28, 31, 52, 55, 75, 80
    Norway, 29, 45, *45*; squat, *46*, 55
London Dumping Convention, 126, 132
lugworm, 33, 59
lumpsucker, 48
Lund University, Sweden, and clean production programme, *101*

M

mackerel, 37, 48, 67, 76, 96
    fishery, 70, 73, 74
magnesium, biological function of, 92
    mining from sea, 109
Markerwaard (IJsselmeer), 65
MARPOL, 132, 136-137
Matthias II, ocean incineration ship, *125*
megrim, 45
mercury, 90, 92, 120
    concentration in seawater, *91*
    in wildlife, 93, 94
merganser, 56, 58
metal, mining, 109
methane, 67, 85
methyl mercury, in body tissues, 93
microflagellate, 29, *29*
military activity, **116-117**
    exercise zones, 112, 116
    shipping, 112, 116, *117*
Moby Dick, Greenpeace ship, *127*
mollusc, 31, 33, 42, 52, 54, 58
    bivalve, 51
Monsanto, pesticide factory, waste discharge, *120*
Mont Louis, sinking, *112*, 113
Moray Firth, Scotland, dolphins in, *95*
motor vehicles, exhaust emissions, 87-88, *88*
    and dioxin production, 100
    lean-burn engines, 88

mud, *14*, 28-29, 45, 51, 54, 59
mud flats, *34*, 62-63, *64*, 66, 69, *69*
    destruction of, 140
mullet, grey, 55
mussel, 29, *49*, 53, 58, 68, *81*
    and toxic algae, 83
    farming, **78**, *78*

N

national parks, 64
    and military exercises, 116, *117*
    habitat destruction, *138*
NATO exercises, 116
Nature and Environment, 121
Netherlands, aerial survey of oil spills, 106
    children and the environment, *129*
    coastal waters, increase in nitrogen, 82
    discharge of sewage sludge, 87
    fishing fleet, 71, 74
    flooding and defences 68, *68*
    gas and oil industry, 104
    Greenpeace in, 120-121
    increase in phytoplankton, 82
    use of lindane, 98
nickel, 90, 92
Nikki, collision, 112
nitrogen, 82, *82*, 85, 86
    agreements on lower levels, 133
    concentration in seawater, *91*
    emissions from power stations, 89
    in atmosphere, 87
    in sea, sources of, 82, 84, 85, 86
    removal from sewage, 87
nitrous oxides, 115
    from chemical industry, 89
    from fossil fuel combustion, 89
    from vehicle exhausts, 87-88, *88*
NL, chemicals factory, 121,122
noise pollution, 140
Nordic Council, 134
Nordstrander Bucht, national park, 64, *138*
North Sea *passim*
    climate, 14, 18
    currents, 14, *14*, *16*, 18
    depth, 14
    drainage basin, *14*
    history of formation, 18-21, *18-21*
    major rivers, *14*
    map of, *14*
    tides, *16*, 18
North Sea conferences, 133-135, 136-137, 140
    and precautionary principle, 133
    decisions legally binding, 129, 134
    Greenpeace actions and influence on, 123-125, *123*, *125*, 127-129
    Scientific Task Force, 133
North Sea Working Group, 121
Norway, 14, 80, *81*
    fish farming in, 75
    fishing fleet, 71, *72*, 73
    magnesium mining, 109
    oil and gas industry, 104, 105, 111
    production of dioxins, 109
    pulp and paper plant, pollution, *100*
Norwegian Trench, 45
nuclear pollution, *120*
nuclear power, **102**
nuclear waste, *102*, *120*, *129*
    dumping of, 133
nuclear weapons, 116
nutrients, 75, **80-89**, *82*, 84, *129*, *129*
    agreement to reduce, 133-134
    from agriculture, 84-85, *85*
    from atmosphere, 87-89
    from sewage, 86-87, *86*
    problems of reduction, 135

O

ocean incineration, 99, *125*, 132, *135*
    agreements to end, 125-126, 133
    areas, 112
    chemicals burned, 125
    Greenpeace reports on, 126
    of halogenated waste, 99-100
octopus, *38*, 55

oil, formation of, 20, *20*
    use in power stations, 87
oil pollution, agreement to reduce, 133
    effect on birds, *111*
    from shipping, 106, 108, 110-111, 113,114
    in Humber estuary, 96
    slicks, 110, *111*
    See also gas and oil industry
oil-based mud, 132,133
OILPOL, 132, 136-137
orca or killer whale, 28, 42-43, *43*
Organization for Economic Co-operation and Development, 138
organohalogens see HHCs
organometallic compounds, 93
Orkney, *81*
Orwell, river, port development, 65
Oslo Commission (OSCOM), 126, 132, 135, 136-137,139-140
Oslo Convention, 121, 136-137
Oslofjord, Norway, discharges in, 83
otter, 28, 96
overfertilization of sea, 82, 84
oxygen, 45, 82, 87
    removal from water by sewage, 86
oyster, 33
oystercatcher, 59

P

PAH (polycyclic aromatic hydrocarbon), 106
Paracas, collision, 112
Paris Commission (PARCOM), 132, 134-135, 136-137, 139-140
PCBs (polychlorinated biphenyls), 90, 95-98, 135
    disposal of, 96
    effect on wildlife, 95, 96
    Greenpeace target, 122
    level in Wadden Sea, 96
    release from dumps, 99
    spillage into sea, *111*
Perintis, sinking, 113
periwinkle, 58, 80
pesticides, *93*
    airborne, 98
    and ocean incineration, 125
    anti-fouling agents, 92, *93*, 106, 132
    HHCs in, 98-99
    loss at sea, 113
    use in fish farms, 75
Phaeocystis, 82, 84
phenol, production and waste of, 103
phosphorus, 82, *82*, 85, 86
    agreement on lower levels, 133
    in sea, *91*
    sources of, 84, 85, 86
phytoplankton, 29, *29*, 30, *82*, 84
    abnormal growth of, 82, 95, 101
    and bioaccumulation, *91*
    and sand and gravel mining, 109
    food web, 59
    natural cycle of, *81*
piddock, 54
Pigment-Chemie, company, 121-122, *121*
pilchard, 28
Piper Alpha, disaster, 110-111, *110*
plaice, 76
    fishery, 71, 74
plankton, 29, *29*, 30, *30*, 31, *32*, *32*, 33, 36, 37, 43, *45*, 51, 52, 54, 55, 58, 59, 70
    effects of cadmium on, 101
    food web, 59
    scientific sampling and monitoring, 72
politics and North Sea, **132-140**
    cost of pollution prevention and control, 138
    fishing regulations, 139
    industry and regulations, 133, 135
    international agreements, 133-134
    North Sea conferences, 123-125, 127-129, 133-135, 140
    regulators and wildlife, 138-139
    regulatory organizations, 132-133, **136-137**
pollack, 29, 37, 58

pollution of North Sea, **90-102**, *94, 95,*
133
  atmospheric see atmosphere
  chemical see chemical pollution
  exhaust emissions see motor vehicle
  licensed discharges, 98
  monitoring difficulties, 101
  oil see oil
  persistence of, *111,* 133-135
  reduction agreements not met, 128
  regulations governing, 132
  reparation for damage, 115
  reverse listing of harmless substances,
    135
  role in seal epidemic, 97, 127, *134*
  routes to sea, 92
  zero discharges and dumping by 2000,
    129, 135
polychlorinated biphenyl see PCB
polycyclic aromatic hydrocarbon see PAH
porpoise, 36, *36,* 77, *91,* 139
  effect of survey explosions on, 105
  effects of fishing on, 77
  harbour, 42
ports, facilities for waste disposal, 114
pout, Norway, 71, 74, 76
power stations,
  and dioxin production, 100
  emissions from, 87, 89
  fluidized bed combustion, 89
  use of catalytic reduction in, 89
prawn, 31, 52, *53*
precautionary principle for protecting
  North Sea, 124-125, 133-134
  implementation of, 127, 129, 134, 140
predation, effect on fish populations, 76
Prince of Wales, North Sea speech, 125
puffin, 28, 41, 56, *56*
  breeding and food shortage, 77

Q R
'Quality Status Report', on North Sea, 123
radioactive waste see nuclear
Ramsar Convention, 136-137, 139
ratfish, 45
ray, 51; eagle, 54; electric, 29
razor shell, 29
razorbill, 28, 41, 56, 110
recycling, in production process, 101
red tide, 83
redshank, 59, 62, 63, 65
regulatory organizations for North Sea,
  132, *136-137*
Rhine, river, estuary, *69,* 93
  Greenpeace actions, *124, 126*
  toxic waste barge on, 120
Rhine Commission, 124, 136-137
Rhizostoma, 30
Rhone-Poulenc, fertilizer manufacturer,
  122
rivers, contamination with HHCs, 98
  discharge of heavy metals into, 94
  pollution of, 90
  pollution routes to sea, 92
rock, *20,* 28-29, 45, 54, 55, 58-59
rocky coasts, *53,* 55, 57
  and oil slicks, 110
rorqual, common, 43
Rotterdam, 93 127
  Greenpeace actions, 120-121, *129*
  pollution survey, 124
  port development, 65
ruff, 63

S
saithe (coley), 37, 76
  fishery, 70-71, 73, 74
salmon, *38,* 48
  farmed, 75, *75*
salt, extraction of, 109
  concentration in seawater, 91
salt marsh, 56, 62-63, *62, 63,* 67, 67, 68-
  69, *69*
  and oil slicks, 110
sand, *16,* 28-29, 45, 51, 59, *63*
  microscopic life in, 59
  mining in North Sea, 109

sand dunes, 62, 66, *66,* 69
sandeel, 33, 36, *36,* 37, 40, 41, 56, 57, 71,
  74, 76, *76,* 139
  collapse of industrial fishery, 77
  effect of sand mining on, 109
sanderling, 59
sandpiper, common, 59
sardine, 67
scallop, 29, 49, 54; great, *50;* queen's, 46
Scheldt, river, 93, 128
Scientific Task Force, 133
scorpionfish, 55
Scotland, 14, *81*
  fish farming in, 75
sea gooseberry, 30, *31, 31*
sea grasses, 62, *62*
sea lamprey, 49
sea lavender, 62, *62*
sea lemon, 54, *54*
sea level, rises in, 67, *67*
sea mouse, 55
sea scorpion, 49
sea slug, *54,* 54
sea-urchin, 28, 29, 33, *38,* 42, 49, 54, 55,
  80
  as keystone species, 34
seabed, 30, *32,* 43, 45, 46, *50,* 51, *51,* 52,
  52, 55, 75, 80
  and heavy metal pollution, 93
  effect of trawling on, 71, 139
  extraction of agricultural lime from, 109
  pollution by oil-based muds, 105
  radioactive waste under, 129
  structure of, *14, 16, 18,* 21, 29, 54
seal, 28, *36,* 38-39, 42, 43, 58, 77, *83,* 117
  and fisheries, **77**
  effects on fish populations, *38*
  epidemic, **97,** 127
  grey, 38, *38,* 39, 42
  harbour (common), 38, *38, 39,* 42
  HHC levels in, 95
  mercury levels in, 94
  PCBs and fertility in, 96
  protection of, 139
seaweed, *38, 48, 58, 58, 86*
  effects of algae on, 80, *81*
Sellafield, UK, radioactive waste from, *120*
Seto Sea, Japan, coastal pollution in, 83
  red tides in, 83
sewage, 86
  dumping, Greenpeace target, 122
  dumping by UK, 87
  effects on sea, 82, 84, 87
  nutrients from, 86-87, *86*
  sludge, contamination of, 83, 87, 93
  disposal of, 86, 87
  treatment of, agreements on, 133
  levels of, 86
  use in agriculture, 87
shag, 56, 57, 58
shark, basking, 45; spurdog, 29, 40; tope,
  29, 40
shelduck, 56
  and military exercises, 116, *117*
shellfish, 28, 29, *32,* 54, 75, 139
  and pollution, 83, 100, 113
Shetland, *81*
  seabird colonies, 77
shingle beaches, 68
shipping, **112-115**
  accidents, 112-114, *112, 114*
  atmospheric pollution from, *115*
  cargo and container ships, *112*
  discharge of oil and chemical waste, 114
  litter from, 114
  military, 112
  navigational aids, *112,* 114
  noise from, 140
  operating standards, 114-115
  pollution damage reparation, *112,* 115
  pollution record, 112
  radioactive pollution from, *120*
shipping lanes, 112, 113
  traffic control, 112, 115
shrimp, 31, 42, 52, 68, 75
silica, 82
Sirius, Greenpeace ship, 122, *122, 123*

Skagerrak, 14, *82*
  chemical weapons in, 116
  currents in, 18
  increase in nitrogen, 82
skate, 29, *48,* 51; long-nosed, 45
slurry, 84-85, *85*
sole, fishery, 71, 74
  lemon, 51
Solent, The, 67
Southern Bight, 14
  increase in nitrogen, 82
Soviet navy, 116
sponge, 28, 29, 54
spoonbill, 57
sprat, 37, 41, 71, 74, 76
starfish, 28, 29, 33, *54,* 55, 78, 80, *81, 106*
  as keystone species, 34
sturgeon, 29
submarines, 116, *117*
Sullom Voe, Shetland, oil terminal, 108,
  108
sulphur, 45
sulphur dioxide, emissions, *115*
Sweden, clean production in, *101*
  fish farming in, 75

T
Tees, river, algae in estuary, 82
  industrial development, 65
Teesside, Greenpeace action, *123*
tellin, 51, 59
tern, 41, 56, 58, 67
  arctic, breeding and food shortage, 77
  common, 57
Texaco Caribbean, sinking, 112
Thames, river, 67, 82, 93
Thames barrier, 69
thongweed, 59
tidal flats, 59
  and global warming, 67
tides, *16, 18,* 58-59, *59*
tin, toxicity of, 92
titanium dioxide, dumping, 121-122, *121*
topshell, *58,* 80
Total Allowable Catch (TAC), 79, 139
toxaphene, 95, 98-99
trawling, 70, *70,* 71, *71,* 139
Trischen, 116, *117*
turbot, fishery, 71

U
United Kingdom,
  and agreement on dumping, 133
  at North Sea conferences, 125, 128-
    129
  colliery spoil dumping, *132*
  'dirty man of Europe', *123*
  Greenpeace target, 123-125, 128-129
  land reclamation, 63
  fishing, 71
  gas and oil extraction, 104, 105
  gravel mining, 109
  Greenpeace in, 120-121
  hazardous items on beaches, 116
  release of PCBs from landfill sites, 99
  sewage sludge dumping, 87,122, 129
  toxic waste dumping, 129
  use of lindane, 98
  vehicle emissions and transport policy,
    88
United Nations' International Maritime
  Organization conventions:
  on marine pollution (MARPOL), 132,
    136-137
  on oil pollution (OILPOL), 132, 136-
    137
  on sea dumping (London Dumping
    Convention), 132, 136-137
United Nations' Environment Programme,
  134

V
Venus shell, 51
Vesta, incineration vessel, 126
Vinca Gorton, collision, 112
Vogelsang, Wadden Sea, 67
Vulcanus II, incineration vessel, 99

W
Wadden Sea, 43, *57,* 64, 67
  effect of trawling on, 71
  gas and oil extraction, 105
  military exercises in, 116-117, *117*
  mussel farming in, 78
  national parks, *138*
  oil spillage, 110
  PCB level experiment, 96
Wadden Sea states, 139
  agreement on pollution reduction, 124
waders, 62-63, 65, *65,* 67, 68, *111*
Wadsy Tanker, dump ship, *122*
Wash, The, 14, 43
  land reclamation, 63
waste,
  auditing, 101-103
  avoidance, 135
  chemical, ocean incineration of, 125
  disposal at sea, 114, *135*
  dumping, licences, 129
  dumping of, 132
  from shipping, 114
  hazardous, *135*
  in water table, 93
  incineration, and dioxin production, 100
  industrial, international agreement to
    end dumping of, 133
  minimization, 103, *103*
  radioactive, dumping of, 129, 133
  reduction of, 101-103, 126, 129
  toxic, 120-121
water meadows, 63
water table, pollution in, 93
wave power, 89
Wedau, incineration vessel, *126*
Weser, river, estuary dredging and heavy
  metal pollution, 93
West Germany, and ocean incineration,
  126
  and PCB levels in fish, 96
  and vehicle emissions, 88
  fishing in North Sea, 71
  gas and oil platforms, 104
  Greenpeace in, 120-121
  ocean incineration of HHCs, 99-100
  release of PCBs from dumps, 99
  use of lindane, 98
wetlands, destruction of, 62-69, *63, 140*
  preservation of, 139
whale, 28, 42, 43; baleen, 43, 51; killer see
  orca; minke, 43; rorqual, stranded, *140*
  effect of survey explosions on, 105
whelk, 29, 49, 80
  common, *50*
  dog, 33, 58
  as predator, *34*
whimbrel, 59
whiting, 58, 71, 76, 80
  blue, fishery, 71, 74, 76
  fishery, 70-71, 73, 74
wind power, 89
witch, 45
wolf fish, *49*
worm, 28, *29, 32, 32,* 33, 45, 55, 58, 62, 65,
  80, 139
  arrow, 31, 33, 36; bristle, 45, 51, 55; cat, 29,
  51, 59; fan, 29, *32, 32;* flat, 54; king rag,
  50; peacock, *46;* rag, 29, 33, 51, 59;
  ribbon, 54; scale, 55; swimming, 31;
  tube, 54
wrack, 58, 59
wrasse, 48, 55, 80
  corkwing, 34

Y Z
Yarrow, dump ship, *123*
Zero 2000, 129
zinc, 90
  biological function of, 92
  concentration in seawater, 91
zooplankton, 29, *29,* 30, *31,* 33
  and toxic algae, 83
  and heavy metal toxicity, 94
  and oil pollution, 106
  natural cycle of, *81*
Zuider Sea see IJsselmeer

ACKNOWLEDGEMENTS

This book would not have been possible without the work of Greenpeace's international North Sea co-ordinator, Andrew Booth, and his assistant, Suzanne Paget, and their predecessors, Alan Pickaver, Monika Griefahn, Ingrid Juetting and Uta Bellion. Among the many others who contributed were Kevin Stairs, James Carr, Nicole Rioux, Lisa Beale, Kieran Mulvaney, Michael Earle, Helene Bours, Andrew Kerr, Topsy Jewell, Lisa Bunin, Beverley Thorpe (GP International), Martin Besieux (GP Belgium), Janus Hillgaard, Ole Lutzen (GP Denmark), Philippe Lequenne (GP France), Henk Kersten, Jan Bylsma, Jikkie Jonkmann, Bas Bruyne, Ron van der Horst (GP Netherlands), Geir Wang-Anderson (GP Norway), Haakan Nordin, Per Rosander (GP Sweden), Tim Birch, Madeleine Cobbing, Paul Horsman, Isabel McCrea, Gillian Glegg (GP UK), and Carsten Redlich and Wolf Wichmann (GP Germany).

The book draws on work carried out for Greenpeace by Paul Johnston and Mark Simmonds (Queen Mary and Westfield College, UK), John Grey (University of Oslo), Bertil Hagerhall (Stockholm), Claus Gulmann (Copenhagen University), Jochen Hanisch, Iciar Oquinena and Michael Günther (Planungsinstitut Küstenregion, Hamburg), Joachim Lohse (Okopol, Hamburg), P. Gerritzen-Rode (University of Amsterdam), D. Ludikhuize (International Centre of Water Studies, Amsterdam), Leo Bass, Harry Hofman, Donald Huisingh, Jo Huisingh, Peter Koppert and Frank Neumann (Erasmus University, Rotterdam), Eddy Somers and Frank Maes (State University of Ghent), Pieter Leroy (University of Nijmegen), Arnaud Lizop and François Veit (Cabinet Lizop, Legal Advocates, Paris), Robin Churchill and John Gibson (University of Wales College of Cardiff), Peter Taylor (PERG, Oxford), Miranda MacQuitty (London), Lynn Barratt and Meg Huby (University of York), Ros Good (UMIST), Peter Kelly and Jacky Karras (University of East Anglia), Alistair Hay (Leeds University) and James Cameron (Gray's Inn and Cambridge University).

Additional thanks to Philip Rothwell, RSPB, UK and Peter Prokosch, WWF–Wattenmeerstelle. Officials involved in the preparations for the Third North Sea Conference also spent time discussing issues, sometimes from a very different viewpoint, but always with the greatest of courtesy.

Thanks are due, too, to Saville of York and MacTel of Nottingham, who supplied Macintosh mice and modems at moments of crisis; to Kate McPhee for her work on the production of the book; Irene Lynch for picture research; Ian Whitelaw for reading the proofs; Liz Somerville for help with Greenpeace photographs; and Jean and Mike Trier for an invaluable index. Finally the author pays tribute to the extraordinary talents of the editor, Lesley Riley, who sustained both the book and himself during a schedule that was frenetic even by publishing standards.

Bibliography

The following volumes were repeatedly consulted during the preparation of this book. *Pollution of the North Sea: An Assessment* (1988), Salomons et al (eds), Springer-Verlag, Berlin; *Environmental Protection of the North Sea* (1988), Newman and Agg (eds) Heinemann, Oxford; *The Waters around the British Isles: Their Conflicting Uses* (1987), Clark, Clarendon Press, Oxford; *Proceedings of the 3rd North Sea Seminar 1989; Distress Signals - Signals from the environment in policy and decision-making* (1989), Ten Hallers-Tjabbes and Bijlsma (eds), Werkgroep Noordzee, Amsterdam; *Northern Europe's Seas*, a report to the Nordic Council's International Conference on the Pollution of the Seas (1989), Aniansson; papers presented at North Sea 2000: Environment and Sea Use Planning Symposium, Heriot-Watt University, Edinburgh,1989; Proceedings of the 5th International Wadden Sea Symposium, Esbjerg, 1986; and of the 6th International Wadden Sea Symposium, List, Sylt; *The Quality Status of the North Sea* (1987), 2nd North Sea conference, Department of the Environment, London; 1990 *Interim Report on the Quality Status of the North Sea* (1990), North Sea Task Force, 3rd North Sea conference, Ministry of Transport and Public Works, The Hague. Information was also gathered from journals including *New Scientist* and *Scientific American*, and government and international regulatory and advisory publications.

Chapter 1: *Atlas of the Seas around the British Isles* (1981), Directorate of Fisheries Research, MAFF, London. *Transport Atlas of the Southern North Sea* (1987), de Ruijter et al, Rijkswaterstaat/Delft hydraulics, The Hague. *A Dynamic Stratigraphy of the British Isles: A Study in Crustal Evolution* (1979), Anderton et al, Allen and Unwin, London. **Chapter 2**: *The Open Sea: The World of the Plankton* (1970), Hardy, Collins, London. *Marine Plankton: A Practical Guide* (1973), Newell and Newell, Hutchinson, London. *Subtidal Ecology* (1987), Wood, Edward Arnold, London. *Key to the Fish of Northern Europe* (1978), Wheeler, Frederick Warne, London. *Seabirds in the North Sea* (1987) Tasker et al, Nature Conservancy Council, Peterborough. *The Handbook of British Mammals* (1977), Corbet, and Southern (eds), Blackwell, Oxford. *Ecological Communities: Conceptual Issues and the Evidence* (1984), Strong et al (eds). Princeton Universty Press, Princeton. **Chapter 3**: 'Management of Wetlands Case Study: The Wadden Sea' (1989), Wolff in *Nature Management and Sustainable Development*, Verwey, W. D. (ed.), IOS, Amsterdam. 'Netherlands: Land Reclamation', Needham, in *European Environmental Handbook* 1987, DocTer International UK, London. **Chapter 4**: *Commercial Fishing Methods* (1986), Sainsbury, Fishing News Books, Farnham; *Dynamics of Marine Fish Populations* (1986), Rothschild, Harvard University Press, Cambridge, Mass. *Marine Ecology and Fisheries* (1975), Cushing, CUP, Cambridge. *Marine Populations: An Essay on Population Regulation and Speciation* (1988), Sinclair, Washington Sea Grant / University of Washington Press, Seattle. *Ecology of Teleost Fishes* (1990), Wootton, Chapman and Hall, London. *Historical and Current State of the Major North Sea Fish Stocks* (1989), Armstrong et al, paper presented at North Sea 2000: Environment and Sea Use PLanning, Heriot-Watt University, Edinburgh 1989. **Chapter 5**: *The invasion of the planktonic algae Chrysochromulina polylepis along the coast of southern Norway in May–June 1988* (1988) Berge et al, Norwegian State Pollution Control Authority Report 339/88, Oslo. **Chapter 6**: *Marine Pollution* (1989), Clark, Clarendon Press, Oxford. *Polychlorierte Biphenyle (PCB): Ein gemeinsamer Bericht des Bundesgesundheitsamtes und des Umweldbundesamtes* (1983), Lorenz and Neumeier bga– Schriften 4/1983, Federal German Government Health Office / Federal Environmental Agency, Bonn. *Polychlorinated biphenyl (PCB) residues in food and human tissues* (1983), MAFF, London. *Cutting Chemical Wastes* (1985), Sarokin et al, Inform, New York. Waste Minimization Opportunity Assessment Manual, Hazardous Waste Engineering Research Laboratory, EPA/625/7–88/003, Cincinnati. *Preventative environmental protection strategy: First results of an experiment in Landskrona, Sweden* (1989), Backman, Huisingh, Persson and Siljebratt, TEM/University of Lund, Sweden. **Chapter 7**: *North Sea Monitor 88.3* (1988), Werkgroep Nordzee, Amsterdam. **Chapter 8**: *Atmospheric Emissions from International Shipping in European Seas* (1987), Laikin, Beijer Institute, York. *Störungen im Wattenmeer: Militarische Übungen* (1989), Wattenmeer International Journal, June 1989. WWF/ IUCN, Husum. **Chapter 10**: *International Conference on Environment and Economics Vols 1 and 2* (1984), background papers, OECD Environmental Directorate, Paris. *UN ECE Seminar on Economic Implications of Low–Waste Technology*, The Hague,October 1989.

Picture acknowledgements

Bryan & Cherry Alexander: 11 top; 70 top; 77 top. **Aquila**: Colin Smith 59; 65 bottom. **Ardea**: Anthony and Elizabeth Bomford 40; Francois Gohier 42; John Daniels 91 top. **Aspect**: 108 top; 134. Henry Ausloos: front cover; 9 bottom left; 38 bottom; 39 top left; 41; 56 top and bottom right; 85 bottom. **Bavaria**: Buchholz 23 bottom, 72 top; 110 inset. **Bildagentur Schuster**: Bramaz 10 top right; Weiβer 24 centre; Jogschies 115 top. **Bilderbeg**: Hans-Jurgen Burkard 63 bottom left, 65 top, 91 bottom left, 93 top left; Ellerbrocke & Schafft 95 top; Wolfgang Volz 104 bottom, 105 top; Frieder Blickle 138 bottom. **Bruce Coleman**: Frans Lanfing 43; Mark Boulton 98 top left. **Diaf**: J. Ch Pratt 21; J. Ch Pratt - D. Pries 25 top and bottom right; L. de Jaeghere 108 bottom. **Anthony Duke** 16 maps. **Mark Edwards/Still Pictures**: half title; 69 bottom; 76 top and centre; 84 bottom; 93 top right and bottom; 94 bottom. **Euromap Ltd**: 15, 60-61. **Fjellanger Wideroe A/S**: 72 bottom. **Geoscience Features Picture Library**: 24 bottom; 57 top; 66 centre; 115 bottom. **Will Giles and Sandra Pond** 34 top and centre; 71; 73; 74 artwork. **Greenpeace**: 16 top; Marriner 39 bottom; Morgan 87 top; Reeve 95 bottom; Landin 97; Grieg 98 top right; Van der Veer 99; Bob Edwards 101; Gleizes 102; Pereira 112 top; Dorreboom 112 bottom,114 top and bottom; Bruyne 117 top; Van Der Veer 118-119; Vennemann 120 top; Dorreboom 120 bottom; Pereira 121 top, 121 bottom; Rajau 122 top; Gleizes 122 bottom; Walker 123 top; Pereira 123 bottom; Vennemann 124 bottom; Geering 124-125; McAllister 125 right; Berndt 126; Goldblatt 127 top; Gibbon 127 bottom left; Dorreboom 127 bottom right; Marccels 128 top; Larsen 128 bottom; Dorreboom 129 top; Neeleman 129 bottom; Richard 133; Greig 135 top; Dorreboom 135 bottom, 140. **Sue Hiscock**: 9 bottom right. **David Hoffman**: 27 top right; 86 ; 89 top and bottom; 90; 94 top; 138 top. **Hunting Technical Services Ltd**: 111 top left. **Hutchison Library**: 10 bottom. **Image Bank**: 22 bottom; Lawrence Fried 79 top. **Images**: 56 bottom left; 96 inset. **John Hilleslon**: Dr Georg Gerster 113 top. **KLM Aerocarto-Schiphol**: 64 top; 66 top left; 68 top. **Laenderpress**: Photri back cover and 11 bottom right; 67; FPG 83; Kreutzfeldt 109; Magr-Panciera 139. **Landscape Only**: 22 top. **Frank Lane**: 79 bottom. **Richard Lewis** 19, 20, 28, 92 artwork.. **Malcolm MacGarvin and Anthony Duke** 35, 136-137 artwork. **Mainbild**: Otto Stadler 23 top left; A. Mosler 32 top. **Mike McQueen/Impact**: 91 bottom right; 132 bottom. **NASA** 12-13. **NCC**: Seabirds at Sea team: 111 top right. **Naturofograferna**: Bo Brannhage 82 top; Klas Rune 85 top; Hans Ring 87 bottom; Klas Rune 88. **NHPA/ Picture Box**: 69 top; 104 top. **NHPA/ Silvestris**: Anke Raotke 130-131. **NHPA/ View As**: Andreas Nygaard 11 bottom left; Per Flathe 100. **NHPA**: L. Campbell **title page**; E. A. Janes 9 top; L. Campbell 27 bottom right; Roy Waller 30 bottom; L. Campbell 38 top, 39 top right; Roy Waller 49 bottom left and right, 52 centre left and right; Jeff Goodman 53 bottom; Nigel Dennis 62; Roger Tidman 63 top; Brian Hawkes 77 centre; L. Campbell 111 bottom. **NRSC**: image produced by RAE Famborough: 14; Dr P. M. Mather, University of Nottingham: 96 main pic. **Oxford Scientific Films**: Doug Allan 6-7; F. Ehrenstrom 8 centre left; George I. Bernard 9 centre right; Peter Parks 29 top and bottom, 30 top, 31 top, centre and bottom; F. Ehrenstrom 32 bottom; Peter Parks 34 bottom; Tony Crabtree 44; Roger Jackman 47; G. I. Bernard 48 top; F. Ehrenstrom 48 centre and bottom; Breck P. Kent 49 top, 50 top left; G. I. Bernard 50 top right; Peter Parks 50 centre; G.I. Bernard 51, 52 bottom, 53 top; Doug Allan 54 top left; G.I. Bernard 54 top right and centre; Colin Milkins 55 top; Roger Jackman 55 bottom. **Katrin Osterlund**, Dept of Physiological Botany, Uppsala University: 81 top. **Planet Earth**: 16 bottom; P. Gasson 25 centre; J. Bracegirdle 25 bottom left; Warren Williams 26 left; Peter Scoones 27 top left; Richard Matthews 27 bottom left; Mark Mattock 32 centre; Linda Pitkin 33; John Lythgoe 36-37; Nicholas Tapp 46 bottom; 52 top; Mark Mattock 57 bottom right; David George 58 bottom left; James King 70 bottom; Brian Coope 105 bottom, 106 top; Alan Stanley 106 centre; Peter Scoones 106 bottom. **Bernard Planterose** 75. **Plymouth Marine Laboratories**: 17, 81 bottom, 84 top. Reprinted with permission from: *The North Sea Satellite Colour Atlas*, ed. J. Aiken, Pergamon Press PLC 1989. **Dr Peter Prokosch**, WWF-Wattenmeerstelle (released under SH/899/57) 117 bottom. **Rijkswaterstaat**, The Hague:132 top. **RSPB**: C. H. Gomersall 63 bottom right. **Scan-Foto A/S**, Oslo: 80 bottom. **Science Photo Library**: George Gomacz 8 top; Michael Marten 8 bottom; 50 bottom; Dr John Green 80 top. **Brian Shuel** 10 centre. **Peter Steinhagen**: 116. **Stief Pictures**: 64 left; 66 top right; 68 bottom. **Tony Stone**: 26 right. **Survival Anglia**: Anthony and Elizabeth Bomford 36 (inset); Rick Price 45; Stone/Deeble 46 top; Anthony and Elizabeth Bomford 57 bottom left. **Swift**: Pete Moore 23 top right, 58 bottom right. **Sygma**: 110 main pic; Alain Dejean 113 bottom. **Svend Tougaard**, Museum of Fishing and Shipping, Esbjerg: 78. **Viewfinder** :107. **Timothy Woodcock**: 23 top centre.